An Assessment of the National Science Foundation's Science and Technology Centers Program

Committee on Science, Engineering, and Public Policy

National Academy of Sciences
National Academy of Engineering
Institute of Medicine

NATIONAL ACADEMY PRESS
Washington, D.C.   1996

NATIONAL ACADEMY PRESS • 2101 Constitution Avenue, N.W. • Washington, D.C. 20418

NOTICE: This volume was produced as part of a project approved by the Governing Board of the National Research Council, whose members are drawn from the councils of the National Academy of Sciences, the National Academy of Engineering, and the Institute of Medicine. It is a result of work done by the Committee on Science, Engineering, and Public Policy (COSEPUP) as augmented, which has authorized its release to the public. This report has been reviewed by a group other than the authors according to procedures approved by COSEPUP and the Report Review Committee.

The **National Academy of Sciences** (NAS) is a private, nonprofit, self-perpetuating society of distinguished scholars engaged in scientific and engineering research, dedicated to the furtherance of science and technology and to their use for the general welfare. Under the authority of the charter granted to it by Congress in 1863, the Academy has a working mandate that calls on it to advise the federal government on scientific and technical matters. Dr. Bruce M. Alberts is president of the NAS.

The **National Academy of Engineering** (NAE) was established in 1964, under the charter of the NAS, as a parallel organization of distinguished engineers. It is autonomous in its administration and in the selection of members, sharing with the NAS its responsibilities for advising the federal government. The National Academy of Engineering also sponsors engineering programs aimed at meeting national needs, encourages education and research, and recognizes the superior achievements of engineers. Dr. William A. Wulf is interim president of the NAE.

The **Institute of Medicine** (IOM) was established in 1970 by the NAS to secure the services of eminent members of appropriate professions in the examination of policy matters pertaining to the health of the public. The Institute acts under the responsibility given to the NAS in its congressional charter to be an adviser to the federal government and, on its own initiative, to identify issues of medical care, research, and education. Dr. Kenneth I. Shine is president of the IOM.

The **Committee on Science, Engineering, and Public Policy** (COSEPUP) is a joint committee of the NAS, the NAE, and the IOM. It includes members of the councils of all three bodies.

This project has been funded with federal funds from the National Science Foundation under contract number STI-9523520. The contents of this publication do not necessarily reflect the views or policies of the National Science Foundation, nor does mention of trade names, commercial products or organizations imply endorsement by the U.S. Government.

Internet Access: This report is available on the National Academy of Sciences Internet host. It may be accessed via World Wide Web at http://www.nas.edu.

Copies of the final report are available from: National Academy Press, 2101 Constitution Avenue, N.W., Box 285, Washington, D.C. 20055; 1-800-624-6242 or 202-334-3313 (in Washington metropolitan area).

Copyright 1996 by the National Academy of Sciences. All rights reserved. This document may be reproduced solely for educational purposes without the written permission of the National Academy of Sciences.

Printed in the United States of America

PANEL TO EVALUATE THE NATIONAL SCIENCE FOUNDATION'S SCIENCE AND TECHNOLOGY CENTERS PROGRAM

WILLIAM F. BRINKMAN, Vice President, Physical Sciences Research, Bell Laboratories, Lucent Technologies
MALCOLM R. BEASLEY, Professor of Applied Physics and Electrical Engineering, Department of Applied Physics, Stanford University
RALPH J. CICERONE, Dean, School of Physical Sciences, University of California at Irvine
GEORGE B. FIELD, Senior Physicist, Smithsonian Astrophysical Observatory and Robert W. Willson Professor of Applied Astronomy, Harvard University
SCOTT E. FRASER, Anna L. Rosen Professor of Biology, California Institute of Technology
ERNEST G. JAWORSKI, Distinguished Science Fellow, Monsanto Company (retired), Ladue, Missouri
LYNN W. JELINSKI, Professor of Engineering and Director of the Center for Advanced Technology in Biotechnology, Cornell University
A. FRANK MAYADAS, Program Manager, Alfred P. Sloan Foundation
JOHN R. RICE, W. Brooks Fortune Professor of Computer Science and Professor of Mathematics, Department of Computer Sciences, Purdue University
J. DAVID ROESSNER, Professor of Public Policy, School of Public Policy, Georgia Institute of Technology
ROLAND W. SCHMITT, Emeritus President of Rensselaer Polytechnic Institute, Rexford, New York
I.M. SINGER, Institute Professor, Massachusetts Institute of Technology
JOHN C. WRIGHT, Professor of Science Education, Institute for Science Education, University of Alabama at Huntsville

Principal Project Staff

DEBORAH D. STINE, Study Director and Associate Director, COSEPUP
SCOTT WEIDMAN, Senior Program Officer
PATRICK P. SEVCIK, Program Assistant
NORMAN GROSSBLATT, Editor

COMMITTEE ON SCIENCE, ENGINEERING, AND PUBLIC POLICY

PHILLIP A. GRIFFITHS (*Chair*), Director, Institute for Advanced Study
BRUCE M. ALBERTS,[*] President, National Academy of Sciences
WILLIAM F. BRINKMAN, Vice President, Physical Sciences Research, Bell Laboratories, Lucent Technologies
DAVID R. CHALLONER, Vice President of Health Affairs, University of Florida
ELLIS B. COWLING, University Distinguished Professor At-Large, North Carolina State University
GERALD P. DINNEEN, Retired Vice President, Science and Technology, Honeywell, Inc.
MILDRED S. DRESSELHAUS, Institute Professor of Electrical Engineering and Physics, Massachusetts Institute of Technology
ALEXANDER H. FLAX, Consultant, Potomac, Maryland
RALPH E. GOMORY, President, Alfred P. Sloan Foundation
M.R.C. GREENWOOD, Chancellor, University of California, Santa Cruz
RUBY P. HEARN, Vice President, The Robert Wood Johnson Foundation
MARIAN KOSHLAND, Professor of Immunology, Department of Molecular and Cell Biology, University of California, Berkeley
THOMAS D. LARSON, Consultant, Lemont, Pennsylvania
WILLIAM A. WULF,[*] Interim President, National Academy of Engineering
DANIEL L. McFADDEN, Director, Department of Economics, University of California, Berkeley
MARY J. OSBORN, Head, Department of Microbiology, University of Connecticut Health Center
KENNETH I. SHINE,[*] President, Institute of Medicine
MORRIS TANENBAUM, Vice President, National Academy of Engineering
WILLIAM JULIUS WILSON, Lucy Flower University Professor of Sociology and Public Policy, University of Chicago

LAWRENCE E. McCRAY, Executive Director

---

[*]Ex officio member.

Preface

The National Science Foundation (NSF) Science and Technology Centers (STCs) program supports 25 academic centers spanning a wide variety of topics, both disciplinary and multidisciplinary. NSF's total program budget for the STC program was some $64 million for fiscal year 1996. The STCs were meant to have an 11-year lifespan. Because the first group of STCs are in their eighth year and they are scheduled to phase out in the ninth year, NSF is now faced with the decision of whether this experimental program should be continued and, if so, in what form.

Thus, in the spring of 1995, NSF requested that the Committee on Science, Engineering, and Public Policy (COSEPUP) of the National Academy of Sciences (NAS), National Academy of Engineering (NAE), and Institute of Medicine (IOM) conduct a study of the STCs program. COSEPUP appointed a panel to carry out this study and this report is the result of the panel's work.

The panel was to review and interpret the data gathered by an outside contractor to NSF (Abt Associates), reach conclusions regarding the progress of the STCs program toward its goals, and make recommendations concerning NSF's future use of the STC mode of support. The study was not to critique the accomplishments of individual STCs except as necessary to draw more general conclusions.

The panel met three times. It heard from the director of NSF, the chair of the NSF advisory committee that oversaw the selection of the current STCs and several directors of the STCs, persons who question the STC approach (primarily because of the concern that it takes funding away from individual investigator awards), and other knowledgeable persons. Beyond that, the panel was able to draw on a number of reports that evaluated the STCs program (described in Chapter 1) and on the knowledge and expertise of the panel members themselves as well as site visit reports and materials provided to panel members by the centers. Each committee member was asked to become familiar with four of the centers so that the panel could have as much information as possible available to it.

The use of Abt Associates as well as the Academy was an experimental effort to have a contractor develop some of the empirical data desired by NSF while maintaining the role of expert guidance from an academy panel. In general, as described in Appendix A, the panel did not view the experiment as successful. Of particular concern was that Abt's initial design was developed without an advisory body and that, as events played out, this panel was brought into the process too late to influence Abt's data collection and analysis. That is of particular concern in that, as indicated by Abt (p. 1–32), "there was no attempt to achieve either 'balance' or an exhaustive range of opinion or judgment in the data collection phase." Furthermore, "it was expected that in most cases, individuals associated with the centers would take the opportunity to 'put their best foot forward,'" and "the survey findings reported . . . should be read with a critical eye, evaluating centers' claims concerning their most important goals, achievements, and impacts against the reader's own understanding of what should be expected of the STC program, given its context and stated purposes. Applying this judgment systematically and distilling the conclusions in the form of concise, definitive, indicator narratives is the role of an expert panel."

In other words, Abt collected primarily positive data from the various participants in the program, as opposed to collecting information and guidance from those outside the program who might have a more balanced and more negative view (such as members of the site visiting committee). In addition, according to Abt, "there is no way to identify retrospectively a proper comparison group for a set of complex scientific enterprises elected in the first place because of their unique characteristics." So there was no time-series analysis, for example, to compare the scientific results obtained by the researchers before and after they became part of an STC. However, several contemporary NSF-funded evaluation studies are dealing with outcomes similar to those projected for the STC program and have managed to construct comparison groups (e.g., ongoing studies by Henderson et al. on academic patents and Feller et al. for COSMOS Corporation on citation measures of EPSCoR and non-EPSCoR researchers).

All these factors affected the panel's belief in and interpretation of the results of the Abt report. As a result, the panel does not recommend that this serial process whereby one contractor carries out data collection and another group reviews the results should be used in the future. The panel did make guarded and selected use of data from the Abt report, but always with supporting evidence (generally provided by the committees that visited the centers on a recurring basis).

Given this concern, it should not be assumed that the panel's work does not itself have flaws. As with most committee-style reports, this report rests primarily on the judgment of its panel–based on the members' own experiences as well as their review and analysis of the visiting committee reports and many other sources of information (described in Chapter 1). Many panel members run centers themselves (although not STCs) and have a good understanding of the benefits and liabilities of such an approach. These judgments are subjective ones and often lack objective data and analysis. Indeed the panel questions whether quantitative measures alone are sufficient to judge whether or not a program is serving research and societal needs. As a result, the panel must rely on its own judgment using the limited valid qualitative and quantitative data that are available.

Although this report represents the work of the panel, it benefited greatly from the support of the staff of COSEPUP, specifically, Deborah Stine, study director and associate director, who managed the project and drafted the report based on panel input; Larry McCray, executive director; Scott Weidman, the initial project director; Norman Grossblatt, editor; and Patrick Sevcik and Dave Amber, who provided administrative support.

The panel also acknowledges with appreciation the presentations made at panel meetings by the following persons:

- Neal Lane, Director, National Science Foundation
- Frank Press, Cecil and Ida Green Senior Fellow, Carnegie Institution
- William Harris, Assistant Director, NSF/Directorate for Mathematical and Physical Sciences
- James Edwards, Executive Officer, NSF/Directorate for Biological Sciences
- David Schindel, Senior Science Advisor, NSF/Office of Science and Technology Infrastructure
- Nathaniel Pitts, Director, NSF/Office of Science and Technology Infrastructure
- Susan Graham, Professor, University of California, Berkeley
- Michael White, Senior Research Scientist, University of Texas, Austin; Director, Center for Synthesis, Growth, and Analysis of Electronic Structures
- Timothy Pickering, Professor, Virginia Polytechnic Institute and State University; Industrial Liaison, Center for High Performance Polymeric Adhesives and Composites
- Bruce Guile, Director, Program Office, National Academy of Engineering
- Aravind Joshi, Professor, University of Pennsylvania; Co-Director and Co-Industrial Liaison, Center for Research in Cognitive Science
- Bernard Sadoulet, Professor, University of California, Berkeley; Director, Center for Particle Astrophysics
- Larry Forney, Professor, Michigan State University; Industrial Liaison, Center for Microbial Ecology
- James Colvard, Visiting Professor, Virginia Polytechnic Institute and State University
- Stephen Fitzsimmons, Vice President, International Operations; Director Center for Science and Technology Policy Studies, Abt Associates, Inc.

- Oren Grad, Associate, Abt Associates, Inc.
- Daniel Kleppner, Professor, Massachusetts Institute of Technology
- Robert Bergman, Professor, University of California, Berkeley

          WILLIAM F. BRINKMAN, *Chair*
     Panel to Evaluate the National Science Foundation's
       Science and Technology Centers Program

Contents

EXECUTIVE SUMMARY		1
1.	**INTRODUCTION**	4
	Charge to Panel	4
	Sources of Information	5
	Report Overview	10
2.	**HAS THE STC PROGRAM ACCOMPLISHED ITS GOALS?**	12
	Observations and Commentary	12
	How Have Other Factors Beyond Those in the Original Solicitations Influenced the Expectations and Operation of the STCs?	12
	How Well Are the STCs Performing Relative to the Research Goal?	13
	How Well Are STCs Performing Relative to the Goal of Education?	16
	How Well Are STCs Performing Relative to the Goal of Knowledge Transfer and Exchange?	17
	Is One Type of STC More Successful Than Another?	19
	What Problems Has the STC Program Faced in Reaching Its Goals?	19
	Findings	20
	Conclusions	20

3. **HOW WELL HAS THE DESIGN OF THE STC PROGRAM WORKED?** 21
   Observations and Commentary 21
   What Was the Design of the STC Program? 21
   How Do STCs Differ from Other Centers and Other Modes
      of Research Support? 23
   Findings 23
   Conclusions 25

4. **HOW WELL HAS THE STC PROGRAM BEEN MANAGED AND EVALUATED?** 26
   Observations and Commentary 26
   Management Issues Encountered by NSF and Individual Centers 26
   Consistent High-Quality Management and Review 28
   NSF's Administrative and Scientific Management of
      the STC Program 29
   STC Program Review 29
   The STC Program Viewed as an Investment by NSF 30
   How Well Does the STC Program Fit with NSF's
      Strategic Plan? 31
   Findings 33
   Conclusions 34

5. **RECOMMENDATIONS** 35

REFERENCES 41

APPENDIXES
A. Review of Abt Report and Comments on the STC Evaluation
   Process 45
B. The National Science Foundation's Science and Technology
   Centers 49
C. Panel and Staff Biographic Information 52
D. Excerpts from Visiting-Committee Reports on the Science
   and Technology Centers 58

Executive Summary

"The National Science Foundation (NSF) established the Science and Technology Research (STC) Centers Program to help maintain U.S. preeminence in science and technology and ensure the requisite pool of scientists with the quality and breadth of experience required to meet the changing needs of science and society—ingredients essential to successful economic competitiveness" (NSF 1989).

"The objectives of the Program are:

- To exploit opportunities in science and technology where the complexity of the research problems or the resources needed to solve these problems require the advantages of scale, duration, and/or equipment and facilities that can only be provided by a campus-based research center.
- To involve students and research scientists and engineers from academia, nonprofit organizations, industry, and Federal laboratories in order to enhance the training and employability of professionals with an awareness of potential applications of scientific discoveries and to provide a mechanism for increasing the transfer of knowledge among sectors of society.
- To provide stable, long-term funding."

Two competitions, in federal fiscal years 1989 and 1991, led to the establishment of 25 science and technology centers in 14 states. The centers span a wide range of science and engineering fields and vary substantially in their responses to the goals of the STC program. When NSF determined that it was appropriate and practical to evaluate how well the STC mode of support had been operating and to recommend adjustments, it asked that the Committee on Science, Engineering, and Public Policy (COSEPUP) of the NAS, the NAE, and the IOM form a panel to evaluate the accomplishments of the NSF STC *program* (not individual centers) against its goals in research, education, and knowledge transfer. This report is the result of the work of the panel charged with that effort. NSF also contracted separately with Abt Associates to gather data about the STC program.

The panel found that most STCs are producing high-quality world-class research that would not have been possible without a center structure and presence, and it found that the design of the STC program has produced an effective means for identifying particularly important scientific problems that require a center mode of support. Many STCs also provide a model for the creative interaction of scientists, engineers, and students in various disciplines and across academic, industry, and other institutional boundaries.

Several other important findings of the committee follow. First, the STC program's vision and themes have evolved over the last 7 years. The NSF has made multidisciplinarity of the research a goal and has placed strong emphasis on K-12 education. Second, although at this stage of their development the individual STCs are largely at the cutting edge of research, the panel found that some of the research subjects are maturing. Finally, it also found that the success of the individual STCs depends critically on the presence of strong scientific and administrative leadership.

On the basis of our studies of the individual centers and the review reports supplied to us, the panel believes that the nation and NSF are getting a good return on their relatively small investment. The total fiscal year 1996 budget for all NSF's modes of support is $3.2 billion. Of the $200 million devoted to all centers at NSF, $60 million goes to the STC program. Thus, less than 2% of NSF's research budget (which does not include any overhead funds for NSF) goes to the STCs. The STCs represent a pilot effort by NSF to achieve a more balanced approach to research problems that are amenable to different modes of support. The panel considers the center approach to be a valuable and necessary tool in NSF's portfolio of support mechanisms. The scope of the center approach is defined by the existence of important research problems that are most amenable to attack by research teams needing the distinctive combination of resources that define centers.

After reviewing the STC program, the panel offers the following recommendations:

1. NSF should continue the STC program.
2. Research and the undergraduate and graduate education linked to it should be the paramount goals of the STC program.
3. In future solicitations, NSF should encourage but not require that proposed STCs be multidisciplinary.
4. The level of funding for the STC program should be maintained to ensure that it retains its strength and vigor.
5. The budget for the STC program should retain a separate identity. More-

EXECUTIVE SUMMARY 3

over, the tradeoff between this program and other NSF activities should be made at the level of the NSF director.

6. STC solicitations should be conducted openly across all fields by NSF as a whole (rather than within specific directorates), and existing STCs should be allowed to compete in this open process.

7. The duration of STC awards should be 10 years. Two periodic solicitations should occur within that period.

8. The differing roles of the NSF Office of Science and Technology Infrastructure and program directorates in the management of STCs are complementary and should be continued.

9. NSF should place greater weight on scientific and administrative leadership in evaluating proposals for STCs and in the periodic reviews of centers.

10. NSF should establish policies allowing center directors to allocate funds and other resources (e.g., staff) both within and among participating institutions, so as to optimize progress toward the center's goals. The limits of this unilateral authority should be clearly defined and procedures to make major reallocations beyond these limits should also be defined.

11. NSF should make every effort needed to coordinate reviews of the centers to avoid redundant data collection and to make previously collected data available to all who may have good reason to be interested.

1

Introduction

The National Science Foundation (NSF) Science and Technology Centers (STC) program was established in 1987 to "fund important basic research and education activities and to encourage technology transfer and innovative approaches to interdisciplinary problems" (NSF 1992). Two competitions, in federal fiscal years 1989 and 1991, led to the establishment of 25 STCs. The centers span a wide range of science and engineering fields and vary substantially in how they have contributed to the goals of the STC program.

The long-term, stable funding that is a hallmark of the STC mode of research support should in principle encourage greater risk-taking and long-term thinking while providing flexibility in setting research directions, initiating projects and creating facilities, and engendering innovative educational ideas. The existence of stable clusters of expertise should also enable the nurturing of technology-transfer pathways from many of the centers. At this point in the evolution of the centers, NSF determined that it was appropriate and practical to evaluate how well the STC mode of support has been operating and to recommend adjustments.

CHARGE TO PANEL

As part of this effort, NSF requested that the Committee on Science, Engineering, and Public Policy (COSEPUP) of the NAS, the NAE, and the IOM form a panel to evaluate the accomplishments of the NSF STC *program* (not individual centers) against its goals in research, education, and knowledge transfer. This report is the result of the work of the panel charged with that effort. In a separate effort, NSF also contracted with Abt Associates to gather data about the STC program.

INTRODUCTION

NSF charged the COSEPUP panel to

- Review and interpret the data gathered by the outside contractor.
- Reach conclusions regarding the progress of the STC program toward its goals.
- Make recommendations concerning NSF's future use of the STC mode of support.

COSEPUP then provided additional guidance to this panel regarding each element of the charge.

With respect to the first charge, the panel not only could consider the data from the contractor, but could consider any information from any other source within the budgetary and time constraints of the study. The panel could also comment on the strengths and limits of the Abt data.

For the second charge, the panel could comment on the STC program's goals, especially as seen in today's emerging changes in the R&D landscape (academic, industrial, and government). The panel could also comment on the appropriateness of the STC mechanism for achieving each of those program goals; this could necessitate comparing the mechanism's strengths and weaknesses with those of the individual investigator and other modes of funding (although a thorough comparison was not intended). And the panel could comment generally on the balance of criteria used in making STC awards (such as scientific merit, relation to national and societal needs, geographic distribution, demographic considerations, and distribution between large and small universities).

As to the third charge, the panel could consider recommendations about the STC program balance—mix of topics, mix of basic versus applied research, interdisciplinary nature of the centers, extent of industry orientation, degree of educational emphasis, and so on—and about management of the STC program without focusing on particular centers (for example, how does NSF structure its oversight?). It could also suggest changes in the mechanism and criteria by which STC proposals are evaluated by NSF, the duration of STC awards, and the mechanism for continuing STC evaluation. Finally, if the panel felt in general that increased or decreased emphasis should be put on the STC program, it could say so; but specific recommendations for program rebalancing could be justified only by a cost-benefit analysis across NSF's entire R&D portfolio.

SOURCES OF INFORMATION

In responding to its charge, the panel was able to use information and guidance from several reports and background documents. These items included the following:

- *Science and Technology Centers: Principles and Guidelines* (NAS 1987)

This report (referred to as the Zare report) by the National Academy of Sciences was written by a committee chaired by Richard Zare. The study was commissioned by NSF as a result of President Reagan's 1987 State of the Union message, in which he called for several initiatives to enhance the nation's economic competitiveness, including the establishment of STCs by federal research agencies. The Zare committee examined the role of NSF in the president's program, the relationship between STCs and other modes of NSF support, essential and desirable features of STCs, mechanisms and criteria for soliciting and selecting proposals to encourage the most-promising ideas, principles and methods of governance (including relationship of STCs to their parent universities, NSF, and their scientific constituencies) and concerns raised within the scientific community by the proposed expansion of the center mode of research. The report provides guidance as to the goals, features, and solicitation and selection criteria for the program, as well as NSF's management of the program and a discussion of the risks of the program.

- *NSF Science and Technology Research Centers: Program Solicitations* (NSF 1988, 1989)

The solicitations describe the features of the STCs, what institutions could submit proposals, requirements for a center director, and selection criteria.

- *Assessment of the National Science Foundation's Engineering Research Centers Program* (National Academy of Engineering 1989)

This report was prepared for NSF to evaluate the mission of the engineering research center (ERC) program. NSF staff drew on existing ERC documents and expertise to develop the proposal solicitation, the review process, the cooperative agreement, etc. for the STC program.

The NAE report concluded that the ERC program was at least as important to engineering schools and industries as when it had first been proposed some 5 years earlier but expressed concern about the adequacy of the ERC program's funding and some aspects of its management.

- *University-Industry Research Centers* (Wesley Cohen, Richard Florida, and W. Richard Goe, Carnegie-Mellon University 1994)

Centers, specialized institutes, and laboratories at academic institutions are not a new idea, and they are widespread. NSF alone has 11 center programs, and many universities have a profusion of centers. For example, this Carnegie-Mellon University report focuses on university-industry research centers (UIRCs) that have been funded by the federal and state government. The national study examined the characteristics and activities of UIRCs, their role in technologic innovation and technology transfer, and the effect of industrial funding on the academic research mission.

- *Alternative Models of Research Performance* (Irwin Feller for U.S. Congress Office of Technology Assessment 1992).

This study evaluates the effects of alternative models of federal support for academic research and of the organization of academic research on research productivity. It evaluates questions related to the effects of teamwork and block-grant funding on creativity, including relative performance (e.g., research productivity and relevance) of the investigator-initiated model and industrial modes of research support.

- *National Science Foundation's Science and Technology Centers: Building an Interdisciplinary Research Program* (National Academy of Public Administration 1995)

This study—requested by the Senate Subcommittee on Housing and Urban Development, Veterans Administration, and Independent Agencies—reviewed the management of the STC program. In conducting its study, NAPA selected five STCs, which were then reviewed by an advisory panel. The study focused on the degree to which the centers exhibited interdisciplinary collaboration, a university base, knowledge transfer to industry, and educational outreach; the value of the center concept of management; the management approaches taken by each center, and NSF's approach to management of the STC program. The study found that STCs are valuable contributions to NSF's research portfolio and to national research goals and are managed in an appropriate manner; that the five STCs are university-based, interdisciplinary, basic-research efforts and good examples for encouraging synergism. It recommended continued funding of the program, continuance of the matrix method of managing the program via the Office of Science and Technology Infrastructure and the responsible directorates, and formation by the National Science Board of a subcommittee to act as a monitoring body and to advise NSF in establishing processes and criteria for management of the program.

- *STC Visiting Committee Site-Visit Reports*

Per NSF requirements, each center was visited several times by an NSF visiting committee of academic and industrial researchers who evaluated its scientific quality and administrative management. These visits included annual site visits during years 1-3 and in-depth reviews in years 3 and 6. Excerpts from these reports (with identifying information removed) are provided in Appendix D.

- *Presentation by STC Directors*

At its first meeting, the panel heard from several directors of STCs. The directors described the attributes of STCs and the strengths and challenges of the program as a whole and their individual centers.

- *An Evaluation of the NSF Science and Technology Center Program* (Abt 1996)

This volume, prepared in parallel with the present report, provides a historical context for the STC program, analysis (bibliometry, patents, etc.), and responses to questionnaires sent to STC directors, university officials, researchers, and graduate students. The key findings from this report are shown in **Box 1-1.**

BOX 1-1
Review of Abt Report Findings on STC Program (Abt 1996)

Principal Research Goals, Achievements and Impacts

1. The science and technology centers are an excellent demonstration of the old axiom that the whole is greater than the sum of the parts.
2. The individual centers have produced significant research achievements in fundamental knowledge and the development of research tools, and have identified a range of downstream impacts of this work.
3. The centers collectively have established a meaning for the Science and Technology concept.
4. A wide variety of approaches to the organization of research and ancillary activities has emerged.
5. The center mechanism, as implemented through the STCs, is seen by survey respondents as enhancing responsiveness, interdisciplinary, and unique approaches to research.

Bibliometric Analysis of Research Performance

1. STC scientists' journal publications were cited 1.69 times as often as the average U.S. academic paper published in SCI-indexed journals.
2. STC papers tend to be published in journals oriented more toward basic than applied research. STC papers in Mathematics and in Chemistry have unusually high representation of industrial organizations in their authorship.

Educational Goals, Achievements and Impacts

1. The individual centers have developed a broad range of educational component and are achieving the objectives established under this program.
2. The STCs have achieved considerable support for their educational programs.
3. According to PIs, the Center context is an especially effective one from which to develop and operate such [educational] initiatives.
4. While all types of educational programs are supported, the most prevalent involve programs for undergraduate students, and outreach programs for underrepresented minorities at the undergraduate level. Precollege educational programs for students, and teacher enhancement programs were the next most frequently emphasized.
5. Support to K-12 teachers, and university and K-12 students were most frequently cited by PIs as their key educational achievements. In many cases, women and underrepresented minority group students were significant beneficiaries of these programs.

continued

BOX 1-1 Continued

6. Educational impacts of these STCs, described by their PIs, include influencing their institution's educational programs, upgrading of science and mathematics in the K-12 sphere, and a series of longer-term improvements in university level science education.

7. There were examples of some impacts of the STCs upon university policies and culture.

Training Support and Job Performance of Graduates of the STCs

1. Almost two-thirds of STC graduates reported having their studies partially or wholly funded by STC-administered research assistantships.

2. STC graduates report being well prepared for their subsequent careers—whether they be in academia, industry or federal laboratory.

3. Certain aspects of the graduates' training can be linked to specific dimensions of job performance. Many graduates continue to participate in cross-disciplinary or industry oriented research in their employment.

Knowledge Transfer Goals, Achievements and Impacts

1. Knowledge transfer activities are seen by the Centers as stimulants to the developments, use, and dissemination of new center research.

2. Most centers focus primarily on traditional academic mechanisms of knowledge transfer.

3. At this stage, centers can demonstrate impressive achievements but have far fewer measurable downstream impacts.

4. Centers have taken measures to be responsive to the needs of the external community, and institutional changes have been made to accommodate knowledge transfer activities.

5. On average, industrial partners consider their affiliations with the STCs to be immensely beneficial.

6. Most industry partners also find many aspects of the center mechanism relevant to their needs.

Patents

1. STC patents have a relatively short technology cycle time (median age of the patents they cite).

2. STC patents are relatively heavily linked to science.

3. STC patents are linked to highly cited earlier patents.

4. A number of STC patents are assigned to private companies.

5. STC research papers are consistently cited by the universe of U.S. patents at a rate 2-4 times higher than the average academic paper.

Program Integration

1. There are synergies among the three major program thrusts of the centers.

continued

> **BOX 1-1 Continued**
>
> 2. There are specific synergies among the scientific activities of the centers.
> 3. There are synergies found in working with industry, other universities, and other Federal or foreign laboratories.
> 4. Centers are achieving fruitful relationships among the faculty, staff, and students, and outside scientists and educators, as various research, educational, and knowledge transfer activities are developed and implemented.
>
> *Management Issues*
>
> 1. There was widespread support among the PIs, advisory board chairs, and deans for the use of a center mechanism as a funding device for the support of fundamental research in the university.
> 2. A number of PIs were very positive in their assessments of the technical support they received through their directorate's technical staff.
> 3. Some PIs believe that NSF has backed off from the STC program, giving it lower status and support in the agency. In particular, the Foundation is seen by some as having given ownership of the program to the directorates and divisions in the face of some vocal critics of the program.
> 4. There is a fairly widespread perceptions among PIs that there are problems with how the STC program is administered; that there is not adequate coordination between OSTI and the NSF directorates with which the centers are affiliated.
> 5. OSTI lost an important program (and possibly policy) mechanism through the cancellation of the STC Advisory Board.
> 6. OSTI and the directorates fail to adequately coordinate the STC site review process.
> 7. There are problems with the volume and frequency of OSTI data requirements and other requests for information; related to this, there were problems with use of these data even when considerable effort have been put into improving the operating data base.
> 8. Selection of the review teams may be excluding some of the most qualified reviewers.

The panel, however, has some concerns about the method of analysis discussed in the report; these concerns are discussed in Appendix A and the preface. For these reasons, only selected data from this report have been used.

REPORT OVERVIEW

The present panel's report focuses on three key questions in response to its charge:

INTRODUCTION

- Has the STC program accomplished its goals?
- How well has the design of the STC program worked?
- How well has the STC program been managed and evaluated?

The findings in response to those questions are provided in Chapters 2, 3, and 4, respectively. The panel used the findings to develop its recommendations, as shown in Chapter 5.

In Appendix A, the panel provides its analysis of the Abt report and comments on the separate contractor/Academy panel approach as a model for future NSF program evaluations. Appendix B lists the STCs. Appendix C provides biographic information on the panel members and staff. Appendix D includes excerpts from the visiting committee reports on the STCs. These reports were a major basis of the panel's recommendations.

2

Has the STC Program Accomplished Its Goals?

This section examines the goals of the National Science Foundation (NSF) Science and Technology Centers (STCs) program and explores the extent to which it has met its goals.

OBSERVATIONS AND COMMENTARY

The STCs were established in response to objectives and features set forward in requests for proposals (RFPs) issued by NSF in 1987 and 1989. These requests were influenced by President Reagan's 1987 State of the Union address and by the Packard-Bromley (White House Science Council 1986) and Zare reports (NAS 1987). The goals and features of the STCs are expressed most clearly in the 1989 solicitation, as shown in **Boxes 2-1**, **2-2**, and **2-3**.

How Have Other Factors Beyond Those in the Original Solicitations Influenced the Expectations and Operation of the STCs?

Recently, NSF has refined the goals for the STC program as expressed in a memorandum from Neal Lane, NSF director, to Alice Rivlin, Office of Management and Budget director, dated September 30, 1994. Briefly, these refined goals are

- Challenging and far-reaching interdisciplinary[1] research problems.
- Knowledge exchange and transfer.
- Education and research training.

---

[1] The panel equates the definition of the terms *interdisciplinary*, *cross-disciplinary*, and *multidisciplinary*. Throughout the remainder of the report, the panel uses its preferred term—*multidisciplinary*.

12

In the original RFP, the multidisciplinary nature of research was not specified. Its inclusion in the goals submitted to OMB constitutes a substantial change, inasmuch as many current centers are not multidisciplinary. In addition, the second and third goals transmitted to OMB seem to have been raised in importance relative to the research goal.

For the purposes of this report, the panel accepts the goals transmitted to OMB as the NSF's most current position. However, the panel believes that

> **BOX 2-1**
> **Over-arching Goal of STC Program as expressed in 1989 Solicitation**
>
> "The National Science Foundation (NSF) established the Science and Technology Research (STC) Centers Program to help maintain U.S. preeminence in science and technology and ensure the requisite pool of scientists with the quality and breadth of experience required to meet the changing needs of science and society—ingredients essential to successful economic competitiveness." (NSF 1989)

- Given that multidisciplinary research was not part of the original solicitation, it cannot be fairly used to judge the existing STC program. (As discussed elsewhere in this report, the panel also feels that this requirement is too restrictive.)
- Similarly, recent changes among goals should not be applied retroactively. The panel believes that research and undergraduate and graduate education linked to that research should be the paramount goals of the STC program. We have evaluated the program in that light.

As with many government programs, the STC program throughout its lifetime has operated under evolving visions and themes. In addition to the modification of the goals, expectations relative to K-12 activities and programs were later imposed on the STC program. Although not a requirement, this addition was strongly influenced by the partnership of the STC program with the NSF Education and Human Resources Directorate (EHR) in which EHR provided additional funds to STCs in exchange for the incorporation of K-12 activities into STC program activities.

How Well Are the STCs Performing Relative to the Research Goal?

To assess the 25 STCs, the panel assigned each member primary responsibility for two centers and secondary responsibility for two others. By reviewing the site-visit reports, brochures, and other documents related to the centers, as well as the Abt report, the panel members were able to familiarize themselves with the accomplishments, impact, and problems of all the centers in a substantive manner. In addition, having panel members with diverse expertise allowed us to obtain views of the STCs from the scientific communities that they serve.

> **BOX 2-2**
> **STC Program Objectives as Expressed in 1989 Solicitation**
>
> "The objectives of the Program are:
>
> " • To exploit opportunities in science and technology where the complexity of the research problems or the resources needed to solve these problems requires the advantages of scale, duration, and/or equipment and facilities that can only be provided by a campus-based research center.
> " • To involve students and research scientists and engineers from academia, nonprofit organizations, industry, and Federal laboratories in order to enhance the training and employability of professionals with an awareness of potential applications of scientific discoveries and to provide a mechanism for increasing the transfer of knowledge among sectors of society.
> " • To provide stable, long-term funding." (NSF 1989)

The analyses indicated that most of the centers have been conducting outstanding research, although in some cases it is difficult to determine the impact of that research in the relatively short time that the STCs have been in operation.

For example, the scientists at the Center for Biological Timing were called an outstanding group of scientists by their site-visit team and have produced an impressive output of scholarly work that has appeared in the top journals in the field. The center succeeded in isolating the first circadian-clock mutant in the mouse and several new ones in plants. Reports on the center indicated that these studies realistically could be accomplished only through center support because of their complexity, their long-term nature, and the unlikelihood of their being supported through traditional investigator-initiated programs. Similarly, the Center for Research in Cognitive Science addresses one of the most important and difficult subjects in all of science and is viewed as having made substantial contributions of very high quality.

Reports on the Center on Synthesis, Growth, and Analysis of Electronic Materials characterize it as an excellent national resource for the fundamental study of the synthesis, growth, and analysis of electronic materials with strong and growing interdisciplinary interactions involving first rate faculty producing high-quality research.

Reports indicate that visitors to the Center for Photoinduced Charge Transfer were uniformly impressed with the consistently high quality of the research projects and with well-thought-out collaborations that led to similar contributions in the field. These accomplishments, according to visitors, resulted in a center that had already matured into a world-class force.

The Center on Superconductivity is widely regarded as having met its primary goal of advancing the fundamental understanding of the new high-temperature superconductors and as having done much better than corresponding centers in other countries.

The panel found that most of the research activities required multi-investigator centers, although not necessarily multidisciplinary ones, for the research to be conducted to its full potential. That is because the problems addressed are too complex to be answered by a single investigator. We believe that several centers, such as the Center for Particle Astrophysics and the Center for Discrete Mathematics, are predominantly in a single discipline.

For example, the Center for Quantized Electronic Structures requires the efforts of materials scientists, physicists, and engineers coordinated in a center approach order to conduct the long-term research on materials synthesis and analysis that will lead to new types of ultrasmall devices that involve the quantum nature of electronic motion. The work conducted on quantum structures at the center is widely respected throughout the world and has drawn many visitors from other universities and industry. One might note, however, that electronic and photonic device science has almost become a field unto itself.

The Center for Clouds, Chemistry, and Climate also needed a center for its activities to understand better the influence of clouds on the earth's atmospheric chemistry. The center includes physicists, chemists, meteorologists, and

---

BOX 2-3
Features of STCs as Expressed in 1989 Solicitation

"STCs should have a unifying research focus involving any field of research supported by the Foundation. The STCs vary in size and exhibit diverse forms of organization, participation, and operation. No single type of center fits the needs of every field. Rather, the size, structure, and operation of the STC is determined by the proposed research.

"While Centers are unique in some respects, each Center must:

" • Be based in an academic institution.
" • Be directed by a scientific or engineering faculty member and integrated into academic programs.
" • Have tangible resource commitments that reflect the priorities of the home institution and other institutional collaborators.
" • Provide a variety of education and research opportunities for students and faculty (e.g., undergraduate and graduate students, postdoctoral researchers, industrial fellows, faculty members from other colleges and universities, including those from institutions without such facilities).
" • Have significant intellectual exchange and substantive resource linkages among various types of institutions (e.g., academic, nonprofit organizations, Federal Government laboratories, industry, Federal, State, and local governments) to facilitate knowledge transfer." (NSF 1989)

oceanographers and has conceived and implemented important field expeditions and made excellent uses for data collected from a National Aeronautics and Space Administration satellite by center scientists. A planned experiment will not only use satellite data but deploy ships and aircraft. Such activities require the coordination of researchers and multi-investigator, multiplatform field experiments that are not possible without a center.

The Center for High-Pressure Research has produced notable and unique scientific and technologic results by taking advantage of stable and centralized funding to create a set of facilities that provided the tools needed for the most advanced exploration and determinations in its field.

The Center for Astrophysical Research in Antarctica requires collaboration by experts in various aspects of astronomical instrumentation operating in a hostile environment. The Center provides infrastructure support that would be very expensive if duplicated by each investigator, as well as the opportunity for investigators to use each other's instruments.

How Well Are STCs Performing Relative to the Goal of Education?

The STC education goal was originally focused on the undergraduate and graduate levels but was later expanded to include K-12 education. There is a wide array of educational activities in the STCs. However, the panel is concerned about the extent to which individual STCs have emphasized K-12 programs. Undergraduate and graduate education is easily coupled to the research endeavor, but that is not necessarily true of K-12 education.

The panel found that all the centers have substantial involvement with undergraduate and graduate students. It is interesting that some of the most important educational activities were not in the typical educational mode (symposia, classes, seminars, and so on) or offered at the centers, but in joint research and information exchanges that were conducted with business and industry (discussed further below).

For example, the Center for High-Pressure Research has produced eight PhDs and three MS graduates, conducted a 10-week undergraduate summer program, and attracted five minority-group students who are pursuing or planning careers in earth sciences or related fields. A unique effort is that of the Center for Particle Astrophysics, which has made an explicit attempt to change the culture of doing science. The center has initiated an in-house forum called "In Balance" to discuss sexual stereotyping, competitiveness, and combining career and family. The forum has sparked similar activities in academic departments, disciplinary societies, and other STCs.

The Center for Molecular Biotechnology has helped to establish the novel Molecular Biotechnology Department at the University of Washington. This department creates an environment and curricula for the development of multidisciplinary scientists, who, through cooperative learning, can enhance the probability of solving complex biologic problems. Visitors believe that the new department will have a significant national impact as a model for research and training.

Many centers have been innovative in K-12 education and found this part of their activities to be enjoyable. For example, the Center for Astrophysical Research in Antarctica has a Space Explorers program in which about 30 high-school students assist in teaching about 2,000 others.

The Center for Clouds, Chemistry, and Climate is participating in a program with the Stephen Birch Aquarium and Museum. There, teachers and research scientists prepare classroom demonstrations and experiments enabling schoolchildren to study climate change. The K-12 activities are supported by San Diego-area industry and other grants, and other NSF support is being sought. Similarly, the Center on Microbial Ecology has developed "The Unseen World," a program that engages teachers in science education and links teachers to resources at the university. Another program, "Science in the City," includes hands-on activities and field trips for high-school students.

The Center for Light Microscope Imaging and Biotechnology has produced a planetarium show, "Journey into the Living Cell," and offers hands-on experiences in the STC teaching laboratory for high-school students, many of whom are members of minority groups. The Center for Engineering Plants for Resistance Against Pathogens developed a computer game called "Germ Wars" and had high-school teachers and students as interns.

Centers also reach out to the general public. For example, the Center for High Pressure Research is developing exhibits and interactive teaching materials with the Long Island Natural Sciences Museum.

How Well Are STCs Performing Relative to the Goal of Knowledge Transfer and Exchange?

A desirable aspect of the STCs is that they are involved in a two-way exchange of knowledge with other researchers, graduate students, businesses, and industry. There is considerable evidence that the STCs as a whole have done an excellent job of disseminating their results whether they are related to basic science, as in the particle-astrophysics center, or more applied fields.

For example, the Southern California Earthquake Center—a collaborative effort of six universities, the US Geological Survey, the Federal Emergency Management Agency, and several state agencies—analyzes many kinds and large volumes of seismic data and theories of earthquakes. It provides its outreach about seismic hazards through publications, and newsletters to regional authorities, industries, and planners.

The Center for Photoinduced Charge Transfer is exploring new modes of knowledge transfer. The center has generated an agreement among the three partners (University of Rochester, Eastman Kodak Company, and Xerox Corporation) that encourages free sharing of ideas and experiments. All research projects are directed jointly by university and industry scientists. Management of Xerox and Kodak have noted a substantial increase in research from their scientists and have seen results translated into products.

The Center for Light Microscope Imaging and Biotechnology interacts productively with established manufacturers of microscopes, and its accomplishments have included establishment of a new company focused on the development of dyes and labeling agents, which has had a major impact on the field.

Similarly, the Center for Ultrafast Optical Science has been exceptionally successful in knowledge transfer. The center has spun off two companies and has had substantial impact on many others. Developments have made their way into commercial products in several larger, more-established companies.

Such active collaborations, especially via joint projects among academic and industry scientists and engineers, are an effective means of increasing face-to-face interactions for exchanging information and transferring knowledge—above that which is achieved solely through scientific publications and conferences. Another benefit of collaborative arrangements is that they seem to improve the ability of graduate students to interact with industrial scientists and engineers. Through those arrangements, students have been able to learn about opportunities and show their abilities to potential employers. The arrangements have also been able to break down the barriers between university and industry and show how there can be a flow of science and personnel between university and industry.

For example, at the Center for High Pressure Research, two companies support center scientists and eight companies have assigned scientists to work at the center's specialized facilities. Visitors have been favorably impressed by university-industry collaboration in a series of systematic tests of the mechanical properties of tungsten carbide.

Relevant portions of the computer industry have provided important and consistent support for the Center for Research on Parallel Computation by making available specialized equipment and involvement of their staff.

The Center for Advanced Cement-Based Materials has an industrial-affiliates program that links the center to 17 industry members who have their own research and development operations; in addition, through a small-business partnership program, the center offers technology-transfer seminars, with approximately 50 persons in attendance from industries that do not have their own research units.

Spinoff companies are an important consequence of knowledge transfer and exchange activities, and they go beyond patents and publications. In some cases, STCs have led to the creation of small businesses to market the technologies developed at the STCs. For example, at the Center for Computation and Visualization of Geometric Structures, a scientist started a company to commercialize a programming system for computer animation.

Is One Type of STC More Successful Than Another?

STCs differ in their degree of multidisciplinarity, their science-technology emphasis, their riskiness, and their facilities. The panel's analysis of the individual centers does not indicate that one type of center is more successful than another in attaining program goals. As will be discussed in Chapter 4 and was discussed in the NAPA report (1995), other factors, in particular center leadership and management, seem to be more important than specific research-center characteristics in making an STC successful.

What Problems Has the STC Program Faced in Reaching Its Goals?

The STC program is not free of problems or shortcomings in meeting its goals. Examination of site-visit reports reveals some general problems and issues. For example, in a rapidly evolving field, it is difficult even for a technically successful center to shift its focus in response to the natural evolution of the field (e.g., from a basic to a more applied orientation or from one scientific emphasis to another). Moreover, any given center might not be well constituted to make such large changes and remain successful. Similarly, a successful center whose original importance and timeliness are derived from long-term research needs of a particular industry, still can find that its programs and the needs of industry drift apart over the course of the lifetime of the center. At least one center, which is universally deemed a technical success, is in the situation that the billion-dollar industry toward whose needs its work is directed exists essentially only outside the United States except for relatively small entrepreneurial firms in niche markets. Clearly, science and technology constitute but one component in economic competitiveness.

Thus, technical success and success in meeting the overarching program goals outlined in Box 2-1 are not necessarily the same. NSF and the US scientific and engineering communities are refining their approach to fostering basic research in support of broad societal goals. The panel believes that further experimentation is called for, but the STC program has been a successful step.

As guidance in designing future steps, the panel offers two observations. First, different parts of science and technology evolve at different rates. No single timeframe can be ideal, but it is certainly the responsibility of NSF to take the long-term, fundamental, and generic point of view. Second, in the modern context, the contributions of science to technology most commonly appear first at the margins, either as one part of a large set of inputs to the technology base of an existing large industry or as the seeds of radical new industries, often expressed through small startup companies. Understanding these dynamics might help in designing future STCs more optimally and astutely.

FINDINGS

The key observations of the panel as to the level of attainment of the STC program goals by individual STCs are as follows:

1. Taken together, the successes in research, knowledge and technology transfer, and education suggest that collectively the centers have achieved the STC program goals.
2. The STC program has undergone shifts of vision and goals. Most recently, that has included a seeming requirement for multidisciplinary research. The panel finds the latter too restrictive. The STC program provides a mechanism to address a wide variety of important problems that might not be multidisciplinary but cannot otherwise be addressed within the NSF structure.
3. Most STCs are producing high-quality world-class research.
4. Most research conducted at STCs would not have been possible without a center structure and presence.
5. In many cases, centers have served as intellectual magnets for the scientists involved.
6. STC programs adequately include undergraduate and graduate students in their activities. Some of them expose students to team research, broader ideas and activities, and broader groups of researchers than typical in traditional department programs.
7. Some centers have interesting, innovative, and vigorous K-12 educational activities.
8. Most centers serve efficiently as two-way conduits between universities and their industrial partners. In general, they perform that function better than traditional departments do.
9. The research characteristics of successful centers vary widely; no single factor determines success.

CONCLUSIONS

Has the STC program attained its goals? The panel believes that the answer is yes. Most of the individual STCs have been successful. They have produced research of high scientific quality with coherent intellectual themes that could be addressed only through center-based research. They have integrated education and training into their activities and have conducted knowledge-transfer and exchange activities with relevant research communities, business and industry, and the public.

3

How Well Has the Design of the STC Program Worked?

The design of the STC program was outlined in the open solicitations for proposals. This open competition mechanism does not just mean that anyone can apply but rather that it is an NSF-wide, directorate-independent competition—not limited to a particular discipline or specialty. NSF appointed a committee to evaluate the proposals. In its evaluation, this committee—known as the STC Advisory Committee—used the criteria from the 1988 solicitation (listed in **Box 3-1**). This chapter explores how well the design has worked relative to the overall goals of the STC program.

OBSERVATIONS AND COMMENTARY

What Was the Design of the STC Program?

The initial design of the STC Program included a number of specific elements. Among these were

- Open competition across NSF directorates.
- Long time scale of grants.
- Multi-investigator research topics.
- Mechanism for knowledge transfer.

The most distinctive aspect of the original competition was its openness. This is one of the few cases in which teams of investigators have been able to propose long-term, large-scale research projects within or across disciplinary boundaries. The openness of the solicitation process has been very successful and has resulted in far more proposals worthy of funding than could be supported in

> **BOX 3-1**
> **Review Criteria Used to Evaluate Proposals for STCs**
>
> - Intrinsic merit of the intellectual focus and research;
> - Research performance competence;
> - Utility or relevance of the research;
> - Appropriateness of the Center approach;
> - Appropriateness of institutional and management plans and arrangements;
> - Effect of the Center on the infrastructure of science and engineering;
> - Quality and appropriateness of the educational and training components of the Center's activities; and
> - Form, appropriateness and strength of linkages, and knowledge transfer efforts to other sectors and groups." (NSF 1988)

the STC program. It is widely believed that many of these proposals led to new research projects but we have no way of examining this issue.

Another key aspect was the importance and timeliness of the proposed research. The centers have addressed problems and opportunities that were important at the time of the competition.

The lifetime of STC funding is 11 years (9 years for the project and 2 years for phasedown). Initial awards were made on February 1 for a duration of 5 years. The cooperative agreements specified that a proposal for renewal was due on July 1 in the third year of support. If the renewal proposal was unsuccessful, the center would have years 4 and 5 of the original award to scale down its activities. If it was successful, the proposal would provide another 5 years of support beginning on the following February 1 for years 4-8. Another renewal proposal was due on July 1 of the sixth year. If it was unsuccessful, the center would scale down in years 7 and 8. If it was successful, another 5 years of funding would start on the following February 1 for years 7-11. Support would scale down in years 10 and 11 by about 20% per year.

At this stage of their development (toward the end of the initial funding period), the panel found that some topics in the original proposals remain timely and others have become mature—in fact, some STCs have evolved from their original goals toward related research. That is partly due to the fast pace of research in many areas of science and partly due to extra risks involved in especially complex research topics. Some will develop better than others.

A further key feature in the original selection criteria was the need for a center-based (multi-investigator) approach, which might or might not have been multidisciplinary. Not all STCs are multidisciplinary. The key common denominator is that the research itself is initiated by multiple investigators and needs a center structure to facilitate it.

How Do STCs Differ from Other Centers and Other Modes of Research Support?

Other NSF centers (e.g., specialized centers, such as earthquake, ecological, and materials centers) and independent centers at universities have multiple investigators and are multidisciplinary. It is important to differentiate how the STC program differs from other NSF modes of support. An attempt has been made in **Table 3-1** (which has been reviewed by NSF staff). The nature of STCs varies widely, so even this table, which describes the many features of a center, does not capture all their attributes. This diversity reflects well the flexibility of the STC program to encompass varied approaches to research.

Note that STCs share with other centers the attributes of timeframe, scale, and, in some cases, multidisciplinarity. The distinguishing characteristic of STCs is the open competition that took place across NSF directorates. Such competition provides the opportunity for the proposal of a wide range of ideas that do not fit into the normal purviews and provinces of NSF directorates or the constraints of the single-investigator mechanism. It also allows an activity to move more quickly into a center mode of funding than in the case of an NSF disciplinary center, such as the Center for Theoretical Physics. And it provides a way of concentrating resources on particularly pressing problems in science or advanced technology that cannot be met with single-investigator support. The open competition across all NSF directorates evoked a particularly intense competition.

FINDINGS

The panel's key findings as to the design of the STC program are as follows:

1. The design of the STC program has produced an effective means for identifying particularly important and timely scientific problems that require a center mode of support. It also provides a model for the creative interaction of scientists, engineers, and students from various disciplines and across academic, industry, and other institutional boundaries. And STCs expose undergraduate and graduate students to the concept of team research, and, in some cases, to multidisciplinary training.

2. STCs are unique in that they are competed for across the entire NSF. STCs share with other types of centers the attributes of time scale and scope necessary to address particularly important problems that cannot realistically be addressed by single investigators. Team-building and coordination are universal characteristics, so excellent scientific leadership is essential. The problems addressed by STCs are both disciplinary and multidisciplinary.

3. In our assessment, STCs are most successful when they are strongly motivated by long-term important problems and opportunities in science and technology; near-term relevance is of secondary importance. That does not mean that relevance to society is unimportant or should be ignored. Rather, the balance

TABLE 3-1 Differentiating Among NSF Modes of Support

	STC	ERC	Specialized Centers (Earthquake; Ecological; Materials)	Industry/ University Coop. Engr. Research Centers	Facility	Group	PI
Competition (**O**pen or **R**estricted to single directorates)	O	R	R	R	R*	R*	R*
Mission (**S**pecific problem/ technology area or **B**road theme)	B	B	B	S	S	S	S
Scope (**D**isciplinary or **M**ultidisciplinary)	D/M	D/M	M	D	D/M	D	D
Timeframe (**L**ong or **S**hort)	L	L	L	L	L	S	S
Scale (**L**arge or **I**ntermediate or **S**mall Number of Persons)	L	L	L	L	L	I	S
Management (**C**oordinated or **N**ot coordinated among research projects)	C	C	C	C	N	C	N
Knowledge Exchange (**L**inked or **N**ot linked)	L	L	L/N	L	N	N	N
External Advisory Committee (**Y**es or **N**o)	Y	Y	Y	Y	Y	N	N
K-12 Education Program (**Y**es or **N**o)	Y/N	Y/N	Y/N	Y/N	N	N	N

NOTE: Keys for table entries are provided in left-hand column.

KEY:
    STC= Science and Technology Center
    ERC= Engineering Research Center
    Coop.= Cooperative
    Engr.= Engineering
    PI= Individual Investigator
    * = There are exceptions, but this is generally true

between scientific and technological importance and relevance to society follows from the unique responsibility of NSF to ensure the long-term vitality of science and engineering in the United States.

4. By design, STCs have long-term support. At this stage of their development, the STCs are largely at the cutting edge of research, but some centers are clearly maturing, as is consistent with the fast pace of research in especially complex research topics.

CONCLUSIONS

The design of the STC program has been successful. On the basis of the panel's review of the 25 centers now in operation, it appears that the most important criterion used by the Advisory Committee for selection of proposals was excellence of science and the appropriateness of the center approach. The STC program constitutes a valuable and complementary mode of NSF support for investigators to develop innovative centers without restriction to a particular directorate or funding restrictions within a particular directorate; thus, it allows resources to flow to the best opportunities. Because centers address complex research questions, require considerable buildup time, and often involve the rallying of a subcommunity of scientists, they need a long-term commitment if they are to be successful. Observing that some centers have already accomplished their goals before the 11-year grant period is over, the panel believes that a shorter award period might be appropriate.

4

How Well Has the STC Program Been Managed and Evaluated?

Management of the STC program nationally is split between NSF's Office of Science, Technology, and Infrastructure (OSTI) and the directorates. OSTI provides administrative management, and the individual directorates provide scientific management. Visiting peer-review committees provide recurring scientific and managerial review of the individual centers. Center directors are, of course, responsible for ensuring that the research, education, and knowledge-transfer activities funded under their proposals are carried out effectively and efficiently. This chapter explores how well this management and evaluation process has worked relative to the overall goals of the STC program.

OBSERVATIONS AND COMMENTARY

The panel relied on several sources of information in evaluating the management of the STC program. Of particular importance were the site-visit reports, discussions with several Center directors, and responses to the surveys of the STC directors conducted by Abt. In addition, the panel reviewed a report of a National Academy of Public Administration (NAPA) study that was conducted in response to a Congressional request (NAPA 1995). The recommendations from that report are summarized in **Box 4-1**.

MANAGEMENT ISSUES ENCOUNTERED BY NSF AND INDIVIDUAL CENTERS

The panel has identified a number of general management issues, some of which are manifested at the NSF level, some at the center level, and some at both levels.

> **BOX 4-1**
> **NAPA Study Elements and Recommendations**
>
> The NAPA study focused on four elements:
>
> - The degree to which the centers exhibited interdisciplinary collaboration, a university base, knowledge transfer to industry, and educational outreach.
> - The value of the center concept of management in science and technology.
> - The management approaches taken by each center.
> - The NSF's approach to the management of the STC program.
>
> The study evaluated five centers to develop its conclusions and recommendations. It recommended that
>
> - NSF, Congress, and the administration recognize the good return on federal government expenditures through the STC program as being in the national interest and worthy of continued support.
> - NSF continue the current matrix method of managing the centers, with funding through the responsible directorates and Office of Polar Programs and coordination through the Office of Science and Technology Infrastructure.
> - The National Science Board (NSB) form a subcommittee, including (as appropriate) representatives of industry and other stakeholders, to
>   — Act as a monitoring body and inform Congress and other stakeholders of the goals, activities, and value of the STC program.
>   — Advise NSF in establishing processes and criteria to assess management of the STCs without adding to the current oversight burden.
>   — Advise NSF in establishing criteria to replace weak center directors or directors who leave.

- The success of centers depends critically on the degree of scientific and administrative abilities of their directors. In the few cases where problems occurred at individual centers, the source of the problem turned out to be the executive leadership of the director.
- Because centers vary widely in their scope, objectives, research foci, appropriate institutional linkages, and so forth, effective control of organizations and programs presents problems for both center directors and NSF program managers.
- Once funded, centers must adhere to the goals and principles that their original proposals espoused. Ensuring that review and monitoring processes are effective can be a problem for NSF managers.

Both NSF and the individual centers have used this accumulated experience to improve their management of the centers.

Consistent High-Quality Management and Review

Center directors must ensure that their centers are not just groups of independent scientists, but rather embody real collaboration. That has not always been the case. For example, site visitors to one center questioned how closely center personnel work together; in another, there was some question as to whether the center funds 35 separate entities or is a true center. Similar concerns have been raised about some parts of two other centers.

Another challenge to leadership is maintaining focus. One center's work potentially affects research on sustainable systems, global warming, hazardous-waste site remediation, and biodiversity. The research problem is important and highly multidisciplinary, but a more-focused activity could result in greater influence on the research community than the research now being conducted, which has resulted in few publications that are merely descriptive.

In another case, a renowned scientist and pioneer in the field first requested that the 3rd-year review be delayed and then in year 4 requested that NSF support be phased out—an unfortunate end for a center whose objectives and approach were ideally suited to the intent of the STC program. The failure was not one of research objectives but rather of a central research tool (a superconducting magnet). It might be that neither the center management nor NSF management fully understood the imperatives of such a major engineering and fabrication challenge. The task of providing the magnet was given to a nonindustrial organization that could not deliver it. Poor management by the STC, NSF, or both appears to have caused the problem. Such mistakes in judgment are unfortunate but sometimes occur at this level of risk-taking. Termination of the center was appropriate.

NSF sometimes provides mixed messages to STCs through its site visits. For example, one center was reviewed three times by NSF; the visits resulted in the appointment of a new director and a vigorous outreach program, particularly in K-12 education, for which it developed software. The NSF staff was particularly constructive in trying to help the center to solve its problems and fulfill its aims. However, although several years ago the NSF reviews said that there was not enough outreach and education, a followup visit recently indicated that there was not enough research. In response to the first site-visit report, almost all the time of the postdoctoral scientists and almost all the resources of the center in the summer months are devoted to K-12 and outreach programs.

One of the centers is in a leadership role in its field. But when this center began, it had some difficulty in managing interinstitutional research, which resulted in warnings from its own advisory committee and a highly critical NSF site review. After a year of provisional funding, the problems were corrected. The same center received conflicting advice on the question of applied research. The center was clearly made up of research scientists who worked on basic materials problems. It was unlikely that they would move into applied research even though the site committee was advising it. At another center, the advisory committee called for in the initial proposal has been inactive, meeting only once in 5 years. From such considerations, we conclude that the overall management of the STCs was very good but can be improved by recognizing the previous pitfalls mentioned and by proper use of the review process.

NSF's Administrative and Scientific Management of the STC Program

When the STC program began, staff from the various NSF directorates was not enthusiastic about the center concept. A separate office within OSTI was created by NSF Director Erich Bloch to oversee the centers. This office designed the program and solicitation and made the original funding decisions. Once the centers came into existence, their management of the program was split between the research directorates and OSTI.

OSTI, with the assistance of the assistant directors and the head of Office of Policy and Planning, functions as the policy-making group. They also design the procedures and formats for proposals, recordkeeping, and so on, but funding and technical management are handled by the various directorates. People from the program directorates rotate through OSTI to learn about the centers and the program.

Although there have been suggestions to shift management to OSTI or to the program directorates, the current arrangement seems preferable if the role of each participant is well understood. Several elements are needed for the STC program to thrive and meet its objectives. First, the research done in each center must be first-rate; this is best ensured by attention of the directorates, who have the best access to high-quality scientific and technical expertise. Second, there needs to be consistency in the management of the program across centers; undoubtedly, differences would emerge if each directorate managed its own centers without reference to those in other directorates. Third, and most important, the budgets for center support must retain a separate identity so that funds can flow to the most promising science independently of directorates. This is particularly important with respect to large grants. The STC program must keep a separate and distinct budgetary identity because the tradeoff of this program with other NSF activities (individual investigator, facilities, etc.) needs to be at the foundation level, with the NSF director responsible for that budgetary decision. The program also needs an advocate outside the individual directorates. For all these reasons, we believe that OSTI or its equivalent must continue to play a strong role.

STC Program Review

The STC program has been reviewed at both the programmatic level and the level of individual centers. At each site visit (annually for years 1-3, in competition for renewal at years 3 and 6, and every 18 months between years 3-6 and after the 6th year), the NSF-selected site-visit team reviews the entire center operation, including the educational and outreach activities. The panel found these reports to be a valuable source of information on the effectiveness of the centers. In several cases where management problems occurred, the NSF periodic site-review process worked well in sorting out the problems and assisted in bringing about solutions. Thus, we believe that NSF site review is important, particularly in the first few years. The value of the peer-review committees is in their expertise in the fields related to the research foci of the centers. However, such groups are less well suited to assess center accomplishments in K-12 outreach programs, because they usually comprise research scientists, not education experts. However, persons from the NSF Education and Human Resources Directorate have participated in some of these visits. In addition, the education expertise of potential site visitors is a factor in selecting teams. In November 1995, two of the STCs made a formal presentation on their educational activities to the Education and Human Resources Advisory Committee.

Center directors express considerable concern that they have been over-reviewed. Several additional reviews (at least of the original 11 STCs) have occurred in recent years, some of which were largely beyond NSF's control. First, the NSF inspector general decided to audit a number of Centers. Second, Congress mandated the study of the STC program by the NAPA. Third, NSF selected the STC program as a pilot project under the Government Performance and Results Act (GPRA), which led to the Abt evaluation and additional site visits. Finally, each STC's external advisory committee (a requirement of the cooperative agreement between each STC's university and NSF) periodically reviews it and advises its directors.

Interviews with center directors reported in the Abt study and this panel's discussions with STC directors support the argument that review is excessive. Moreover, directors point out that many of the reviews ask for the same information, causing them to wonder how much of the information is used or even readily available for use.

The committee does not endorse the recommendations by NAPA for deeper NSB involvement in the review process. NSB—through its membership, committee structure, and procedures—is well equipped to review the broad goals, strategies, and priorities of programs, but it is not the appropriate forum in which to assess how well each activity is performing. Thus, a review of our assessment of the entire STC program is an appropriate subject for NSB attention, but its direct involvement in the reviews of individual centers is inappropriate, except to hear and use such reviews.

THE STC PROGRAM VIEWED AS AN INVESTMENT BY NSF

The total fiscal year 1996 budget for all of NSF's modes of support is $3.2 billion; a breakdown is shown in **Table 4-1**. Of the $200 million devoted to all

TABLE 4-1 NSF 1996 Budget by Modes of Support (in millions)

	Amount	Proportion of Research Budget
Research projects (individual-investigator and group grants)	$1,700	53%
Facilities (large multiuser facilities)	$700	22%
Education & Training (K-12, undergraduate, graduate, postdoctorate, and underrepresented groups, such as women and minorities)	$600	19%
Centers (ERCs, earthquake, ecology, materials, minority, industry/university cooperatives, mass spectrometry)	$140	4%
STCs	$60	2%

centers at NSF, $60 million goes to the STC program. Thus, less than 2% of NSF's research budget (which does not include any overhead funds for NSF) goes to the STCs. That is a very small investment, compared with the other modes of support shown in Table 4-1 and **Figure 4-1**.

The panel considers the STCs as examples of a mode of support that allows particular types of research problems to be addressed that otherwise would not be. If research problems are regarded as arrayed along a spectrum, with some problems well-suited to individual-investigator modes of inquiry, others to a center mode, and others to a facility mode, STCs emerge as one mode of support that helps balance the NSF portfolio of funding instruments. STCs constitute an experimental beginning of an effort to achieve balance. No evidence suggests that the limits of research problems best suited to center-like modes of support have been reached.

How Well Does the STC Program Fit With NSF's Strategic Plan?

NSF's current strategic plan puts forth a mission, NSF's vision, several long-range goals, and core strategies. The long-range goals, in abbreviated form, are

- To enable the United States to uphold a position of world leadership in all aspects of science, mathematics, and engineering.
- To promote the discovery, integration, dissemination, and employment of new knowledge in service to society.
- To achieve excellence in US science, mathematics, engineering, and technology education at all levels.

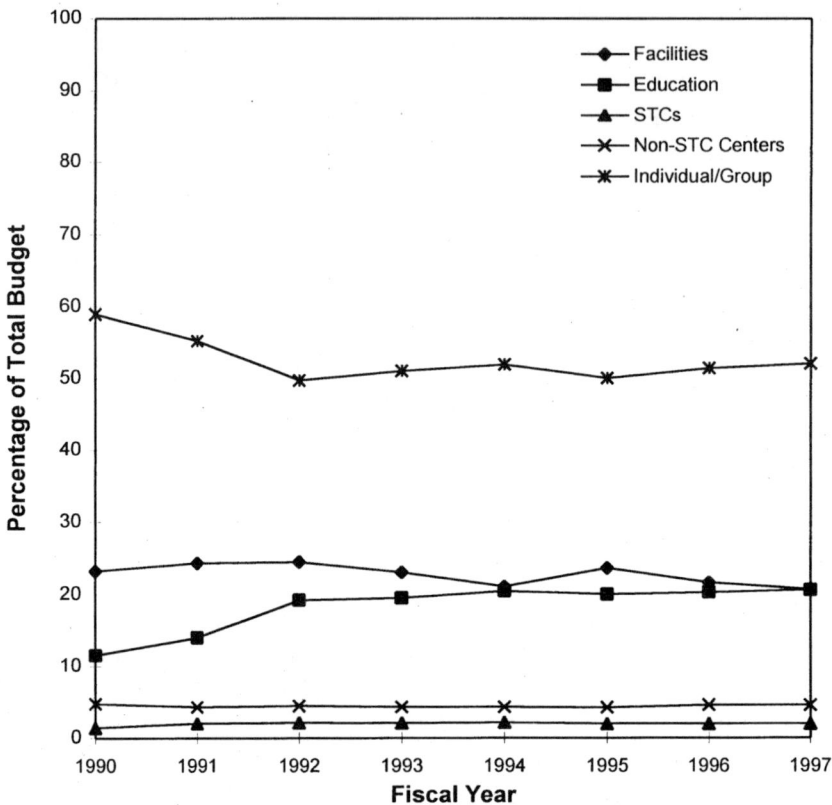

FIGURE 4-1 NSF research budget by modes of support, 1990-1997.

The core strategies, abbreviated, are

- To develop intellectual capital.
- To strengthen the physical infrastructure.
- To integrate research and education.
- To promote partnerships.

Clearly, the goals and strategies demand a variety of approaches. The individual-investigator mode is certainly an important and proven mode in meeting some of the goals and strategies, but other modes are needed to meet goals that require multidisciplinary and multi-investigator efforts. Science and engineering research and education have thrived in a variety of venues, and NSF must offer a rich array of modes for the best results. In fact, the NSF strategic plan states explicitly that NSF "encourages flexibility in the methods used to promote the progress of science and its benefits to society." Thus, the STC program fits NSF's strategic plan well.

The STC program was based on the idea that subjects to be addressed through the center mode of support should be identified by the research community. This panel endorses that approach. The role of NSF management is to state the general program goals clearly. The panel urges that any changes in goals or in priorities among them be clearly articulated and communicated to the centers and be reflected in new solicitations.

FINDINGS

The panel's key observations as to management and evaluation in the STC program are as follows:

1. The success of individual centers depends critically on having strong scientific and administrative leadership.

2. The nation and NSF are making a good investment in the STC program, gaining considerable leverage in research with a relatively small part of the total NSF budget. The STCs constitute an effort by NSF to achieve a balanced approach to research problems amenable to different modes of support. The panel considers the center approach to be a valuable and necessary tool in NSF's portfolio of support mechanisms. The limitations of the center approach are defined by the existence of research problems that are most amenable to attack by research teams with the combination of resources that defines centers.

3. The STC program needs two kinds of management: administrative and scientific. Centralizing the program in one NSF directorate is difficult because no single office has sufficient scientific expertise. The original program announcement and selection needs to come from a central office to open up the process fully and allow funding to move to the best centers across all fields of science.

4. The reviews of centers formally required by NSF are frequent but not excessive. However, the burden placed on some centers is excessive because of additional reviews–such as those of the NSF inspector general, NAPA, and the GPRA pilot program. The normal review process facilitated the solution to several management problems when they occurred in specific centers; it plays a valuable role. To the extent that multiple layers of review seek the same information from centers, NSF should try to coordinate reviews so as to avoid redundant data collection and to make data previously collected available to all who have good reason to be interested.

5. Because NSF has most recently appeared to place equal weight on research, education, and knowledge exchange goals, some center directors have as well.

6. There is some concern that resources devoted to K-12 activities compete with the time, energy, and funds devoted to research and that, if centers desire to pursue these activities, they not overwhelm the core mission of centers—research and education at the university level.

7. The STC program fits in well with NSF's overall strategic plan.

CONCLUSIONS

The STC program is making a major contribution to the nation's scientific and engineering research for a relatively small investment. Management of individual STCs and NSF's management of the overall program are critical for their success. NSF's mechanism for managing the program seems to be working well, although attention should be paid to balancing centralized administrative (budgetary) management against decentralized scientific management. Furthermore, program goals and priorities should be clearly articulated and communicated and closely linked to overall NSF priorities. Of particular concern are the equal emphasis that NSF now places on the three program goals, with the inclusion of K-12 education as an addition to these program goals, and the evolutionary manner in which these priorities have shifted. At the individual center level, success requires a combination of a high degree of scientific leadership and enthusiasm and unusual administrative skills.

5

Recommendations

The panel offers the following recommendations based on its review of the STC program:

1. NSF should continue the STC program.

NSF support for centers is based on the premise that some compelling scientific questions can best be addressed through the long-term, coordinated efforts of teams of researchers. STCs are unique among NSF centers in that they create opportunities for research that are unconstrained by the NSF directorate structure. The program also provides a means for the focusing of substantial resources on frontier problems within the directorates; these problems might require facilities beyond the scope of individual grants and require coordination through center funding. The driving forces for creating centers are the scientific importance of such research and the resulting societal benefits.

2. Research and the undergraduate and graduate education linked to it should be the paramount goals of the STC program.

The primary functions of universities are education and research. The STC program's primary goal must be long-term research. Associated with that are undergraduate and graduate education and knowledge exchange with other institutions and industry. Education and the exchange of knowledge make research valuable to society by making sure the knowledge is used.

Recently, the NSF has placed considerable emphasis on K-12 educational programs in the STCs. In our view, these activities should be undertaken in ways that are natural extensions of particular centers' activities. In general, they should be of secondary importance. If K-12 activities are to be continued, they should not be prescribed. Centers should be allowed to judge the appropriateness of topics and opportunities for K-12 outreach. In addition, a process for evaluating their educational effectiveness needs to be put in place.

3. In future solicitations, NSF should encourage but not require that proposed STCs be multidisciplinary.

The STC program has undergone shifts of vision and goals. Most recently, the changes included an implied requirement for multidisciplinary research. That is too restrictive. The STC program provides a mechanism to address a wide variety of important problems that are not otherwise possible within the NSF structure. They may or may not be multidisciplinary.

The panel endorses the view of the Zare report (NAS 1987):

> [Science and technology centers'] primary goal is to exploit opportunities in science where the complexity of the research problems or the resources needed to solve these problems require the advantages of scale, duration, or facilities that can be provided only by the center mode of research. . . . Interdisciplinary research, although essential for the solution of many problems, should be pursued only when there is a demonstrated need or opportunity, not because of current fashions or the enhanced likelihood of funding.

Multidisciplinary research should not be a requirement in future STC solicitations. As indicated in the original solicitation shown in Box 2-1, the STC program should "help maintain U.S. preeminence in science and technology and ensure the requisite pool of scientists with the quality and breadth of experience required to meet the changing needs of science and society." The benefit of this view is shown in the current STCs—most of which have provided unique problem-solving abilities but that are both disciplinary and multidisciplinary.

4. The level of funding for the STC program should be maintained to ensure that it retains its strength and vigor.

When the STC program began, there was some concern that substantial funds would be diverted from individual-investigator activities. The panel found that not to be the case. The fraction of NSF's research budget devoted to the STC program is very small—about 2%. Yet, this is the only NSF program that provides open competition across all directorates. In our view, the investment is modest and has paid large dividends. Sufficient funding is available to support a variety of exciting endeavors without taking a great deal of funding from other NSF programs.

The NSF director and the National Science Board need to ensure that, at the margin, investments in each of the various NSF modes of support—projects, centers, facilities, education, and training—are matched appropriately to the nature of the problems faced in the fields of science and engineering supported by NSF. In other words, an additional dollar of support for individual investigators or facilities should be as well matched to the problems that are best addressed by that mode as an additional dollar of support for centers is matched to the problems best addressed through centers.

In addition, concerns were expressed at the time of program initiation that the program would result in projects that funded weak researchers and would not receive adequate peer review; on the basis of the site-visit reports and the publication records of the STC researchers, this appears not to be the case. However, in the first two rounds, there were far more appropriate and excellent proposals than could be funded. The site visit report, in particular, repeatedly noted that the STC they visited supports some very strong and excellent investigators. Moreover, individual centers are monitored closely, perhaps overreviewed rather than underreviewed. Although not all STCs have been successful and not all fields of research will retain their timeliness over the full 11 years, the STC program is important because it meets an important need in NSF's modes of support.

5. The budget for the STC program should retain a separate identity. Moreover, the tradeoff between this program and other NSF activities should be made at the level of the NSF director.

To fulfill its roles, the STC program must keep a separate and distinct budgetary identity, and the tradeoff of this program with other NSF activities needs to be at the foundation level, with the NSF director responsible for that budgetary decision. The program also needs an advocate outside the individual directorates. We believe that Office of Science and Technology Infrastructure or its equivalent must continue to play a strong role.

6. STC solicitations should be conducted openly across all fields by NSF as a whole (rather than within specific directorates), and existing STCs should be allowed to compete in this open process.

The STC program allows scientists to propose research ideas that are multidisciplinary or that do not fit well within the modes of support provided by the current NSF directorates. The panel believes that it is important to support promising initiatives in science and engineering that cross disciplinary lines or that are otherwise frustrated by the NSF structure. Solicitations for the STC program should be managed by NSF as a whole so that competition can take place across all program directorates and funds can be provided for the fields of greatest opportunity.

Some have suggested that existing STCs not be allowed to recompete, inasmuch as they have already had their chance to explore a given research subject for a long period and others should be provided an opportunity. But some STC subjects are still important to explore and of great relevance, and existing STCs have advantages over new competitors in having had the experience of making an STC work, having the infrastructure in place to support an STC, having collaborative groups already functional, and having management and advisory teams trained and functional. Although that might give them an advantage during the review process over those who are just sending in proposals, the panel believes that existing STCs should be allowed to compete. They should have neither advantages nor penalties in this process. It is the purpose of this open competition of allowing funds to go to the best ideas and not to allow the kind of permanence that other centers seem to achieve.

7. The duration of STC awards should be 10 years. Two periodic solicitations should occur within that period.

The dynamic nature of science and technology requires that NSF accept funding requests on a regular cycle. The panel recommends that solicitation be conducted every 4 years with two solicitations in the 10-year span of each grant. This would allow two years of ramping up of the new programs and two years simultaneous phaseout of the old programs. In addition, it would spread the opportunities for funding and the necessary intense solicitation and review process more evenly over time.

The current duration of 11 years (9 years plus 2 years for phasedown) is working well but could be shortened by 1 year. If future competitions occur in year 8 now or year 7 in the future and decisions are announced in year 9 now (or year 8 in the future), it would allow for the current 2-year phasedown period (and a 2-year startup).

8. The differing roles of Office of Science and Technology Infrastructure and program directorates in the management of STCs are complementary and should be continued.

No program office can cover the enormous scope of scientific activities in the STC program. But there needs to be consistency in the goals of the program and in the procedures used for proposals, recordkeeping, reporting, oversight, and so forth. The best way to achieve that is to continue managing the STC program as it is now, taking advantage of the complementary roles of OSTI and the program directorates.

9. NSF should place greater weight on scientific and administrative leadership in evaluating proposals for STCs and in the periodic reviews of centers.

Scientific leadership is the key to a successful center, especially if it is managing highly complex, multi-investigator, multi-institutional research.

A number of the STCs have had problems in scientific leadership or administrative management. NSF has responded to the concerns well; the centers that have had difficulty were able to make corrections within a short period. The National Academy of Public Administration report indicated that NSF does not have a clear method for removing or replacing an STC director and that the National Science Board should be involved in developing such a process. But the panel believes that the existing process works. We have several examples of cases in which the review process did its job. However, the panel also believes that the evaluation of the scientific leadership and administrative management of a center should be made an explicit part of the site-visit review process, which is not now the case. NSF should immediately respond to adverse site-visit reports and provide the university an opportunity to make corrections; this has generally been done, but it should be more explicit in NSF procedures.

There has been a tendency on NSF's part to conduct a full review of a center even for an administrative problem that does not involve the center's scientific work. That is unnecessary and unduly burdensome. The follow-up reviews should focus on only the aspects of the STC that are of concern.

10. **NSF should establish policies allowing center directors to allocate funds and other resources (e.g., staff) both within and among participating institutions, so as to optimize progress toward the center's goals. The limits of this unilateral authority should be clearly defined and procedures to make major reallocations beyond these limits should also be defined.**

In some cases, agreements have been made among the institutions setting up a center for a specific allocation of funds to the host institutions. That makes it difficult for centers to shift resources over time to the subjects of greatest need. NSF must require that these agreements—which are often set at the proposal stage—include a process to change allocations. Center directors should have the freedom to move resources to where they are most needed. Given the length of the award for centers, one cannot expect an original allocation always to make sense 5-10 years later.

11. **NSF should make every effort needed to coordinate reviews of the centers to avoid redundant data collection and to make previously collected data available to all who have good reason to be interested.**

There has been considerable concern that the STCs are overreviewed. The panel believes that the 3-year cycle of the NSF is appropriate. However, the addition of the other reviews (by NAPA, NSF, and the Inspector General, and as a

Government Performance) and the fact that advisory committees sometime act like review committees has led to excessive reviews. The NSF needs to assist in controlling this process by avoiding redundancy of reviews and by making data previously collected available to all legitimate claimants.

References

Abt Associates. 1996. *An Evaluation of the NSF Science and Technology Centers (STC) Program*. Cambridge, MA.

Cohen, W., R. Florida, and W.R. Goe. 1994. *University-Industry Research Centers*. Carnegie-Mellon University. Pittsburgh, PA.

Feller, Irwin. 1992. *Alternative Models of Research Performance*. U.S. Congress Office of Technology Assessment. Washington, DC.

Feller, Irwin, Robert K. Yin, and Cheryl Sattler. 1995. *Plan to Evaluate the National Science Foundation's Experimental Program to Stimulate Competitive Research* (Revised, October 1995). Ongoing study prepared for the National Science Foundation. COSMOS Corporation. Bethesda, MD.

Henderson, Rebecca, Adam Jaffe, and Manuel Trajtenberg. 1995. *Universities as a Source of Commercial Technology: A Detailed Analysis of University Patenting 1965-1988*. National Bureau for Economic Research Working Paper 5068. Washington, DC.

NAE (National Academy of Engineering). 1989. *Assessment of the National Science Foundation's Engineering Research Centers Program*. Washington, DC.

NAPA (National Academy of Public Administration). 1995. *National Science Foundation's Science and Technology Centers: Building an Interdisciplinary Research Program*. Washington, DC.

NAS (National Academy of Sciences). 1987. *Science and Technology Centers: Principles and Guidelines*. Washington, DC.

NSF (National Science Foundation). 1988. *NSF Science and Technology Research Centers: Program Solicitation*. Washington, DC.

NSF (National Science Foundation). 1989. *NSF Science and Technology Research Centers: Program Solicitation*. Washington, DC.

NSF (National Science Foundation). 1992. *NSF Science and Technology Centers*, NSF report 92-104 (August). Washington, DC.

White House Science Council. 1986. *Report of the White House Science Council Panel on the Health of U.S. Colleges and Universities*, Executive Office of the President, Office of Science and Technology Policy (February). Washington, DC.

APPENDIXES

APPENDIX A

Review of Abt Report and Comments on the STC Evaluation Process

The preface discusses the panel's general belief's about the Abt report. Two major issues are addressed here:

- Review and interpretation of the data gathered and reported on by Abt Associates.
- Comments on the STC evaluation process.

To establish the context for our discussion of these issues, we first review the objectives of the Abt evaluation of the STCs, as stated in its report, and the relationship that NSF expected the Abt evaluation to have with the COSEPUP panel's objectives.

According to volume I of the evaluation of the STC program by Abt, the evaluation had three objectives: "1) to provide relevant and timely information to NSF decisionmakers considering whether or not to continue support of the STC program as presently constituted; 2) to document whether or not the STC program's research centers were, in the aggregate, accomplishing their research, education, and knowledge transfer objectives consistent with the original rationale for the STC program; and 3) to provide inputs to a pilot evaluation process under the Government Performance and Results Act (GPRA). . . . The study sought to identify aspects uniquely attributable to the center mechanism of operation."

Abt envisioned the role of an independent expert panel as follows (volume I, page ix): "In preparing the study design for the present evaluation, we outlined a rationale for incorporating a series of qualitative dimensions of performance in the evaluation of a fundamental research program, and for the use of an expert panel to assess the quality of the program's research and other accomplishments on the basis of structured, qualitative data. However, such a panel was not included in the study."

The COSEPUP panel was specifically asked by NSF to

> (1) review and interpret the data gathered by an outside contractor (Abt Associates); (2) reach conclusions regarding progress the STC program has made toward its goals; and (3) make recommendations concerning NSF's future use of the STC mode of support. [memorandum from Larry McCray to COSEPUP Panel to Evaluate the NSF's Science and Technology Centers Program, September 22, 1995].

NSF viewed the STC evaluation process as a possible model for future evaluations of programs of this type:

> The NSF program evaluation staff considers the STC Program to be an experiment, and is interested in the results of the COSEPUP and Abt Associates evaluations both for the resultant insights into the program and also for insights into the art of program evaluation. [COSEPUP proposal to NSF, September 12, 1995, p. 2].

REVIEW AND INTERPRETATION OF THE ABT REPORT

The Abt report is based on a historical review of the STC program, analysis of secondary data, bibliometric and patent analyses, and surveys of populations associated with the centers. The historical review developed information on changes in basic program structure, program goals and changes, eligibility, guidelines, criteria for review, review procedures, and management policies and practices. Secondary data included information from OSTI and NSF databases and OSTI copy files on the 25 STCs with respect to center funding, staff, and students; copies of the original grant-award jackets; files on the third-year reviews required for all centers and conducted via site visits by peer researchers. Bibliometric analyses were used to study the amount and quality of publications by STC participants, patterns of coauthorship of the publications across institutions, patterns of citations of the publications, and research foci of publications (applied versus basic). Patent analyses were used to measure the influence, science linkage, and technologic currency of STC-based patents.

Eight groups of persons associated with STCs were surveyed

- Principal investigators (all).
- STC advisory board chairs (all).
- University deans or provosts (all).
- Industry or federal laboratory representatives (three persons identified by each STC).
- Educational-outreach collaborators (three persons identified by each STC).
- STC administrators (all).
- STC graduate-school alumni (all).
- Job supervisors of graduate school (all).

APPENDIX A   47

The report offers a large amount of descriptive material about individual centers and the STC program itself. Particularly illuminating and rich are the numerous quotations from the groups surveyed. Collectively, the quotations document qualitatively the wide variations in research foci, educational outreach, and knowledge transfer among centers, and the strengths and weaknesses of the center concept and its administration by NSF. Also useful are the individual center profiles, which contain time-series data on sources of support, graduate and undergraduate educational activity, educational outreach-activity, and interaction with industry and federal agencies.

However, the report suffers from serious shortcomings that limit its usefulness for evaluating the STC program. First, the primary data on the achievements and impacts of each center consist of self-reports by persons who, because they are direct or indirect beneficiaries of the center, do not have an independent perspective. Thus, the report assumes the tone of an advocacy document much like a legal brief, rather than an objective program evaluation. Readers do not fully recognize until they are deep into the report that there are no data from groups that might have independent or even negative views about the centers.

Abt surveyed various participants in the STC program with the expectation that, "in most cases, individuals associated with the centers would take the opportunity to 'put their best foot forward'" (Abt volume II, page 1-32). An expert panel, Abt assumed, would incorporate this positive bias in its deliberations and achieve a balanced set of conclusions.

In justifying that approach, Abt stated (page 1-32) that "there is no source impartial or 'neutral' opinion about the centers; individuals who are knowledgeable in depth about a center are likely to be either professional collaborators or competitors of the center." The COSEPUP panel rejects that position because it implies that the initial STC-proposal review process, the periodic site-visit process, and indeed the entire peer-review process are invalidated.

Second, and similarly, the study was not designed to elicit data from comparison groups with which the importance of the achievements and impacts identified in the surveys can be judged. The achievements and impacts described stand in isolation; there is no objective basis on which to assess the extent to which they can be attributed to the features characteristic of centers or whether they might have occurred in the absence of centers. Readers of the report must rely on the perceptions of principal investigators and others who have substantial stakes in the success of the centers with which they are affiliated.

Bibliometric analyses potentially offer a basis for judging the quality and quantity of publications generated by center participants and comparable groups, but the potential was not taken full advantage of. Rather than comparing the publications of center participants with their own prior records (one indicator of the influence of center participation on publication activity and patterns) or with the contemporaneous records of comparable researchers in non-STC environments (another appropriate indicator of the quality, quantity, and nature of center publications), Abt chose to compare center output with that of all researchers publishing in similar journals in the same fields. As a result, few clear conclusions can be drawn about the effect of center participation on researchers' publication records—and thus about the effect of centers themselves on the quantity and quality of research output in their fields. Abt also could have examined how center participation might have changed the foci of investigators' research.

Third, Abt did not carry out any analyses of the relationships among the several key sources of data or among categories of respondents that produced the same types of data. Doing so would have constituted an important check on the validity of the achievement and impact data. In the former case, the principal investigators' self-reports of the centers' most-important achievements and impacts could have been compared with the available NSF site-visit reports for each center and with the center-profile data on graduate and undergraduate education, educational outreach, and industry activity. In the latter case, the perceptions of principal investigators could have been compared with those of advisory board chairs. Consistency would have increased confidence in the results, and discrepancies could have been explored with data from other sources.

COMMENTS ON THE ABT-COSEPUP PROCESS AS A MODEL FOR FUTURE EVALUATIONS

If NSF wishes to have the benefits of different program-evaluation methods in deciding the fate of programs, such as the STC program, it should either (1) follow a carefully designed and reviewed agency-generated overall evaluation plan, contract for separate independent studies that use alternative methods or strategies (one of which might be a peer-review panel), and synthesize the resulting data through inhouse evaluation or (2) support an evaluation by a single professional evaluator who maintains full control over the evaluation design (which could include a peer-review component).

The panel strongly recommends against NSF's use of a process like the one used in the STC program evaluation as a model for future program evaluations. As the process has evolved in this case, the panel has been placed in the position of acting as a check against the positive biases of the Abt report. At the same time, the timing of the process was such that Abt was unable to incorporate comments from the panel about questions addressed in the evaluation or on the evaluation design itself. This failure is both untenable and unacceptable.

APPENDIX B

The National Science Foundation's Science and Technology Centers[1]

1. Center for Clouds, Chemistry, and Climate
   University of California, San Diego, California
2. Southern California Earthquake Center
   University of Southern California, Los Angeles, California
3. Center for Advanced Cement-Based Materials
   Northwestern University, Evanston, Illinois
4. Center for Research on Parallel Computation
   Rice University, Houston, Texas
5. Center for Computer Graphics and Scientific Visualization
   Cornell University, Ithaca, New York
6. Center for Research in Cognitive Science
   University of Pennsylvania, Philadelphia, Pennsylvania
7. Center for Biological Timing
   University of Virginia, Charlottesville, Virginia
8. Center for Analysis and Prediction of Storms
   University of Oklahoma, Norman, Oklahoma
9. Center for Magnetic Resonance Technology for Basic Biological Research
   University of Illinois, Urbana-Champaign, Illinois
10. Center for Discrete Mathematics and Theoretical Computer Science
    Rutgers University, Piscataway, New Jersey
11. Center for Advanced Liquid Crystalline Optical Materials
    Kent State University, Kent, Ohio

---

[1]See map on p. 51 for locations.

12. Center for Superconductivity
    University of Illinois, Urbana, Illinois
13. Center for High-Pressure Research
    State University of New York, Stony Brook, New York
14. Center for Quantized Electronic Structures
    University of California, Santa Barbara, California
15. Center for Astrophysical Research in Antarctica
    University of Chicago, Yerkes Observatory, Williams Bay, Wisconsin
16. Center for Particle Astrophysics
    University of California, Berkeley, California
17. Center for Light Microscope Imaging and Biotechnology
    Carnegie Mellon University, Pittsburgh, Pennsylvania
18. Center for Ultrafast Optical Science
    University of Michigan, Ann Arbor, Michigan
19. Center for Molecular Biotechnology
    University of Washington, Seattle, Washington
20. Center for Engineering Plants for Resistance Against Pathogens
    University of California, Davis, California
21. Center for Microbial Ecology
    Michigan State University, East Lansing, Michigan
22. Center for High Performance Polymeric Adhesives and Composites
    Virginia Polytechnic Institute and State University, Blacksburg, Virginia
23. Center for Computation and Visualization of Geometric Structures
    University of Minnesota, Minneapolis, Minnesota
24. Center for Synthesis, Growth, and Analysis of Electronic Materials
    University of Texas, Austin, Texas
25. Center for Photoinduced Charge Transfer
    University of Rochester, Rochester, New York

APPENDIX B

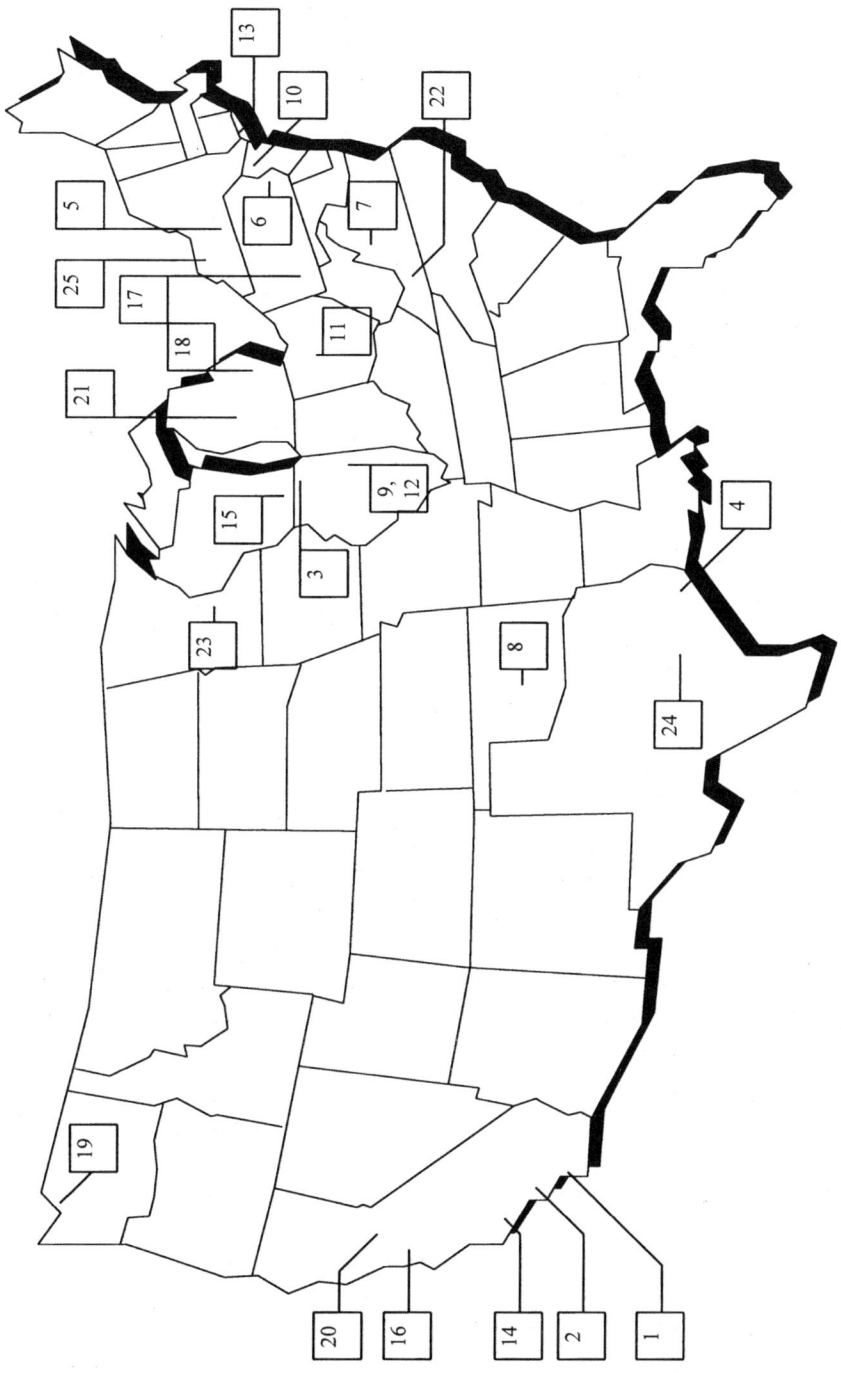

APPENDIX C

Panel and Staff Biographic Information

PANEL

William F. Brinkman is vice president of physical sciences research for Lucent Technologies, formerly AT&T Bell Laboratories; he held the same position at AT&T Bell Laboratories. He was vice president of the Sandia National Laboratories in 1984-1987, director of the Chemical Physics Research Laboratory in 1981-1984, head of the Infrared Physics and Electronics Research Department of Bell Laboratories in 1972-1974, and a resident fellow at Oxford University in 1965-1966. Dr. Brinkman received his BS (1960), MS (1962), and PhD (1965) in physics from the University of Missouri at Columbia. He received an honorary DHL from the same institution in 1987. He is a member of the National Academy of Sciences.

Malcolm R. Beasley has been professor of applied physics and electrical engineering (by courtesy) at Stanford University since 1980; he was associate professor from 1974 to 1980. Dr. Beasley was a resident fellow of engineering and applied physics at Harvard University from 1967 to 1969 and then assistant professor and associate professor from 1969 to 1974. He received his BE (1962) in engineering physics and PhD (1968) in physics from Cornell University. He is a member of the National Academy of Sciences.

Ralph J. Cicerone is professor in the earth system science department and dean of the School of Physical Sciences of the University of California, Irvine. Previously, he was a senior scientist and director of the Atmospheric Chemistry Division of the National Center for Atmospheric Research (1980-1989). Dr. Cicerone was also a research chemist at Scripps Institution of Oceanography at the University of California, San Diego (1979-1981), and the University of Michigan at Ann Arbor (1970-1978). Dr. Cicerone received his SB (1965) from the Massachusetts Institute of Technology, and his MS (1967) and PhD (1970) in electrical engineering and physics from the University of Illinois. He is a fellow of the American Association for the Advancement of Science, the American Meteorological Society, the American Geophysical Union, and a member of the National Academy of Sciences.

George B. Field is senior physicist at the Smithsonian Astrophysical Observatory and Robert W. Willson Professor of Applied Astronomy at Harvard University. Formerly, he was a professor of astronomy at Princeton and the University of California, Berkeley. He was director of the Harvard-Smithsonian Center for Astrophysics from 1973 to 1982. He received his BS in physics from Massachusetts Institute of Technology (MIT) in 1951 and his PhD in astronomy from Princeton in 1955. He has received the Joseph Henry medal of the Smithsonian Institution and several awards from the National Aeronautics and Space Administration. He is a member of the American Academy of Arts and Sciences and the National Academy of Sciences.

Scott E. Fraser is the Anna L. Rosen Professor of Biology and serves as the Director of the Biological Imaging Center of the Beckman Institute at Caltech. Previously, he was an assistant professor (1980-1986), associate professor (1986 to 1990) and full professor (1990-1991) at the University of California, Irvine, where he served as the chair of the Department of Physiology and Biophysics (1989-1991), and the assistant director of the Developmental Biology Center (1985-1991). He received his BS (1976) from Harvey Mudd College and his PhD (1979) from Johns Hopkins University. He is a fellow of the American Association for the Advancement of Science.

Ernest G. Jaworski was distinguished science fellow with the Monsanto Company from 1970 to 1991. He has been an assistant biochemist (1949-1952), a research biochemist (1952-1954), a research group leader (1954-1960), a scientist (1960-1962), and a senior scientist (1962-1970). He received his BChem (1948) from the University of Minnesota and his MS (1950) and PhD (1952) in biochemistry from Oregon State University. He is a fellow of the American Association for the Advancement of Science.

Lynn W. Jelinski is professor of engineering and director of the Center for Advanced Technology in Biotechnology at Cornell University. At AT&T Bell Laboratories, she was a staff fellow in biophysics (1978-1980), a member of the technical staff in chemistry (1980-1984), head of polymer chemistry (1984-1985), and head of biophysics (1985-1992). Dr. Jelinski received her BS (1971) from

Duke University and her PhD (1976) in chemistry from the University of Hawaii. She was a fellow in chemistry at Johns Hopkins University in 1976-1977 and a fellow at the National Institutes of Health in 1977-1978.

A. Frank Mayadas is program manager at the Alfred P. Sloan Foundation. Before joining Sloan, he held various positions at IBM: member of the research staff of Watson Research Center (1965-1971), manager of the thin-film and metal group (1971-1975), manager of memory and storage research (1975-1977), manager of the technical planning staff (1977-1979), manager of storage systems and technology at the San Jose California Research Laboratory (1979-1981), director of technical planning and controls, director of the San Jose Research Laboratory, and director of the IBM Almaden Research Laboratory and Research Division vice president; IBM director and secretary, IBM Management Committee. Dr. Mayadas received his MetE (1961) from the Colorado School of Mines and his PhD (1966) from Cornell University. He is a fellow of the Institute of Electrical and Electronics Engineers.

John R. Rice is W. Brooks Fortune Professor of Computer Sciences and professor of mathematics at Purdue University. He was a research fellow in mathematics at the National Bureau of Standards in 1959-1960 and a senior research mathematician at General Motors Research Laboratories in 1960-1964. Dr. Rice received his BS (1954) and MS (1956) from the Oklahoma State University and his PhD (1959) in mathematics from the California Institute of Technology. He is a member of the National Academy of Engineering, a fellow of the American Association for the Advancement of Science, and a fellow of the Association of Computing Machinery.

J. David Roessner is professor at the School of Public Policy of the Georgia Institute of Technology and program manager for technology policy at SRI International. Before going to the institute, he held positions at Hewlett-Packard (1964-1965), the Bureau of Social Science Research, Inc. (1970-1973), the National Science Foundation (1977-1978), and the Solar Energy Research Institute (1978-1980). At the Georgia Institute of Technology, he has been acting director (1983-1984 and 1988-1989) and director (1990-1992) of the technology and science policy program, and interim director (1990-1992), associate director (1992-1995), and director of graduate studies (1994-1995) of the School of Public Policy. He has been with SRI International since September 1995. Dr. Roessner received BS and MS degrees in electrical engineering from Brown University (1962) and Stanford University (1964), respectively. He received an MA (1967) and PhD (1970) in science, technology, and public policy from Case Western Reserve University.

Roland W. Schmitt is president emeritus of Rensselaer Polytechnic Institute and senior vice president (retired) of science and technology with General Electric. He has also been a member and chairman of the National Science Board, the governing body of the National Science Foundation. He is a member of the National Academy of Engineering, a foreign member of the Royal Swedish Academy of Engineering Sciences, and a foreign associate of the Engineering Academy of Japan. Dr. Schmitt received his BA and BS (1947) and MA (1948) from the University of Texas and his PhD (1951) in physics from Rice University. He is also a fellow of the American Physical Society, the American Association for the Advancement of Science, the American Academy of Arts and Sciences, and the Institute of Electrical and Electronics Engineers. He is currently chairman of the board of governors of the American Institute of Physics and chairman-elect of the Council of Scientific Society Presidents. He holds nine honorary doctorates and Rice University's Distinguished Alumni Award.

I.M. Singer is institute professor at the Massachusetts Institute of Technology (MIT). He has held positions as Moore instructor of mathematics at MIT (1950-1952), assistant professor at the University of California, Los Angeles (1952-1954), visiting assistant professor at Columbia University (1954-1955), visiting member of the Institute of Advanced Study at Princeton University (1956), professor of mathematics at MIT (1956-1970), and Norbert Wiener professor (1970-1979), visiting professor (1977-1979) and professor of mathematics at the University of California, Berkeley (1979-1983), Miller Professor of mathematics at the University of California, Berkeley, and John D. MacArthur professor of mathematics (1983-1987). Dr. Singer received his BS (1944) from the University of Michigan and his MS (1948) and PhD (1950) in mathematics from the University of Chicago. He is a member of the National Academy of Sciences, the American Academy of Arts and Sciences, the American Mathematical Society (vice president, 1970-1972), the American Physical Society, and the American Philosophical Society.

John C. Wright is professor at the Institute for Science Education of the University of Alabama at Huntsville. He has been an associate professor (1957-1959) and professor (1959-1961, chairman) at West Virginia Wesleyan College (1957-1959), an assistant program director for precollege programs at NSF (1964-1965), dean of the college of arts and sciences of Northern Arizona University (1965-1970), dean of the college of arts and sciences at W. University (1970-1974), vice chancellor for academic affairs of the W. Board of Regents (1974-1978), and president of the University of Alabama (1978-1988). Dr. Wright received his BS from the University of West Virginia and his PhD in chemistry from the University of Illinois.

PANEL STAFF

Lawrence E. McCray is director of the National Research Council's Policy Division and executive director of the Committee on Science, Engineering, and Public Policy (COSEPUP). Dr. McCray held positions in the US Environmental Protection Agency, the US Regulatory Council, and the Office of Management and Budget before coming to the Academies in 1981. He has directed academy studies in carcinogenic risk assessment, export controls, nuclear winter, and federal science budgeting. A Fulbright scholar in 1968, he received the Schattschneider Award in 1972 from the American Political Science Association for the best dissertation in American government and politics. In 1987, he received the National Research Council Staff Award.

Deborah D. Stine is study director and associate director of COSEPUP. Dr. Stine has been working on various projects throughout the academy complex since 1989. She received a National Research Council group award for her first study, for COSEPUP, on policy implications of greenhouse warming, and a Commission on Life Sciences staff citation for her work in risk assessment and management. She holds a bachelor's degree in mechanical and environmental engineering from the University of California, Irvine, an MBA, and a PhD in public administration, specializing in policy analysis, from American University. She received an International Mitchell Prize Young Scholar's Award for her personal research in International Environmental Decisionmaking. Prior to joining the Academies, she worked for the State of Texas and the Chemical Manufacturers Association.

Scott T. Weidman is senior program office for this study and also serves as director of the National Research Council's Army Research Laboratory Technical Assessment Board. Dr. Weidman joined the National Research Council in 1989 with the Board on Mathematical Sciences and moved in 1992 to the Board on Chemical Sciences and Technology. In late 1995, he assumed his current position directing a new board that will provide annual technical assessments of the R&D performed at the Army Research Laboratory. He earned bachelor's degrees in mathematics and materials science from Northwestern University and an MS and a PhD in applied mathematics from the University of Virginia. Before joining the National Research Council, Dr. Weidman worked for General Electric Corp., General Accident Insurance Co., Exxon Research and Engineering Co., and MRJ, Inc.

Patrick P. Sevcik is the program assistant for the Panel and COSEPUP. Before his work at the National Research Council, Mr. Sevcik was an assistant program officer with the International Republican Institute (IRI) from 1990 to 1993 working primarily in Central and Eastern Europe. He has held positions at the White

House in the Office of Political Affairs (1989-1990) and on Capitol Hill (1987-1988) in the office of Representative John DioGuardi (R-NY). During this time, Mr. Sevcik also held concurrent positions in several Slovak-American organizations. Mr. Sevcik holds a BA in international affairs, with an emphasis on Soviet and Eastern European studies, from the George Washington University. He has also studied Russian language and culture at the Leningrad Polytechnic Institute in former Leningrad, USSR.

APPENDIX D

Excerpts from Visiting-Committee Reports on the Science and Technology Centers

As discussed in the preface and Chapter 1 of the report, many of the panel's judgments about the centers are based on its reviews of visiting committee reports. At each site visit (annually for years 1-3, for 3- and 6-year renewal competitions, and every 18 months between years 3-6 and after the 6th year), an NSF-selected visiting committee reviews the operation of a center. The visiting committee includes about 5-10 persons who have expertise in the subject of its center's research.

The panel found these reports to be a valuable source of information. Each site-visit report was generally organized according to the goals of the program (research, education and outreach, and knowledge transfer) and provided an overall evaluation of the program (generally in terms of whether funding should be continued for the center).

The reports are long so it is not possible to include them in their entirety or to note every interesting point made in them. However, we provide here excerpts from the visiting-committee reports so that readers can gain some context for the panel's qualitative analysis of the STC program and resulting conclusions and recommendations.

According to the panel's agreement with NSF, these reports can be quoted only if the centers are not identified and if information that could identify the centers is not included.

APPENDIX D

TABLE D-1 General Comments

STC	General Comments
A	The Center has assembled an impressive group of scientists and students to pursue the stated goals of [the center]. The Site Visit Panel (SVP) finds that the group is highly qualified to carry out the proposed research program. It has the breadth and expertise necessary to address critical questions relative to [the center's specialty]. Recent [center] activities have galvanized interactions among individuals at [the] participating institutions. The Director has demonstrated extraordinary capabilities for organizing and leading the [center] team. The entire team is highly motivated and extremely dedicated to the goals of the center. The SVP strongly recommends without any reservations that the [center] program be supported at the level requested.
B	The Center approach to this problem is proving to be effective by fostering cooperation between disciplinary groups which attack the same scientific problem from different, complementary directions. An example of this synergy occurred in response to [an incident] where participants in the [center] mounted an effective field effort. They made several significant discoveries which will result in a new paradigm for evaluating [the center's specialty]. The site review unanimously recommends that the [center] be awarded full support as requested with no significant reservations.
C	The site visit team is very satisfied with the way the Center has responded to concerns expressed in the last site visit report. The team members are also impressed by the commitment and enthusiasm shown by the Center members. The concern raised in the [previous] site visit report dated . . . , was the need for more emphasis on research [in a subject area] and the adoption of the research approach in studying the physical phenomenon. . . . Our observation is that the Center has effectively implemented the suggested approach and developed new projects addressing [the subject area].
D	In summary, the Center is an effective national Science and Technology Center that is conducting world-class research. . . : [the center] has contributed significantly to the growing use of [the center's technology]; and [the center] programs have had a meaningful impact on K-12 and undergraduate education, with substantial involvement of women and underrepresented minorities.
E	The Center is conducting exciting research that has significant strategic value. The Center was criticized in past site visits as lacking a strategic plan and supporting little collaborative work. To the credit of the participants, they have responded to this criticism and are embarking on many new collaborative projects. This progress shows that there is clear value added by being an STC. However, more effort needs to be expended by the STC leadership to articulating the goals of the STC, to formulating a stronger strategic plan, and to prioritizing projects within the context of this plan.
F	The following report on the site visit, which generally confirmed and reinforced this enthusiastic reaction to the proposal, includes some minor criticism that do not affect the overall highly positive response of the site visitors. We strongly recommend that the renewal proposal be approved. The renewal proposal articulates a bold and ambitious vision of [the center's subject area] as a strong interdisciplinary approach to the [center's] sciences. . . . The faculty are an

TABLE D-1 *Continued*

STC	General Comments
	outstanding group of distinguished scientists. The ongoing and proposed research is of excellent quality, although some areas have not been fully developed as yet. The educational component is strong and diverse. Linkages to industry and governmental organizations are many, as are academic contacts. The added value of the STC is impressive in all the areas mentioned. Weaknesses include the slow development of interdisciplinary research and education in [one of the disciplines] and between that [subject area] and [others]. However, this imbalance has begun to be rectified and will be given attention in the future. A final weakness is the lack of space and the postponement of new space allocations. None of these weaknesses reduce the enthusiasm of our recommendation for renewal of this grant.
G	Not only would this type of inherently risky and long-term venture would be unlikely to be supported through traditional investigator-initiated grant programs, but it also requires the synergistic efforts of a group of scientists to succeed. In general, the Site Visit Team felt that the Center was moving in the right direction, and that weaknesses would be overcome as the Center matures.
H	The Site Visit Panel commends the performance of [center] scientists during the first half of the project and recommends without reservation the continuation of the Center. The Center Mode of Support for [this center] has allowed for the development, starting from scratch, of a [new] system built with extensive collaboration between . . . scientists [in one field] and [another] using state-of-the-art . . . technology. The long term support enabled the careful consideration and unprecedented documentation of [the science] which would not be possible in the more normal research environment of short term funding. This newly developed tool has risen to a state of prominence equal and, in some cases higher, than other . . . systems developed over the period of decades by virtue of its user friendliness and . . . In order to ensure the greatest probability for the success of the project, the Site Visit Panel recommends that, during the remaining five years. [The center] should adjust their mix of personnel to reflect a greater percentage of full time research associates and post-doctoral employees.
I	An independent review team should be sent to [the center] facilities to examine the prospects for completion of the current project. The STC team at . . . needs to bring on a strong competent "executive science officer" to be the day-to-day front-end for the program with, initially, a strong focus on ensuring timely completion of the [device].
J	As emphasized earlier in the report, the visitor remain impressed by the excellence of the center's activities. For that reason, we unequivocally RECOMMEND continuation of the center for the entire remainder for the 11 year term. The previous site visit report emphasizes the need for feedback and evaluation of operations and programs. Little progress was made. Management believes evaluation is too costly and would serve little purpose. This will make it difficult, if not impossible, to convince funding sources beyond NSF to support [the center]. Little has changed in the evaluation of [the center] workshops. They may or may not be evaluated through participant questionnaires. Yet both last year and this, with regard to workshops, members of the site visit team heard positive anecdotes from attendees, but also anecdotes on the lack of scientific

TABLE D-1 *Continued*

STC	General Comments
	communication and disorganized local arrangements. Lack of uniform and objective feedback will prevent the center from improving its product. Unfortunately, these two steps will leave the center far removed from where it should be at this point. The emerging leadership is handicapped by lack of a plan a list of specific long-term [the center] goals. These elements are needed to define the center and establish its continued existence. The management team has not yet made adequate progress in articulating specific goals and focusing the center's mission towards them.

K	This center is unique.
	    Nowhere else in the U.S. is there a research Center of this size and productivity focusing on [this subject].
	    The site visit team is very impressed with the Center's technical, educational and outreach achievements and planned activities. The site team is particularly impressed with the leadership and the management, and [the center's] successful coordination of [several] universities into a synergistic, highly interactive Center.
	    The site team strongly recommends renewal for a five year period at the requested level.

L	The site visit has convinced us that the STC is working well. In general, their research remains world-class. Their industrial and educational outreach activities are features which could not be achieved in small group settings. Center researchers remain optimistic and energetic, at a time when individual investigators are having a hard time sustaining programs in [the center's subject area].
	    Does the center mode of funding provide added value? For the STCs the answer is certainly yes and it was clear from talking to the students that the ties between institutions that were generated by the center were in a large number of cases crucial to the success of the individual student and postdoc projects.
	    We feel that the center mode is working for the STCs and has provided an institution that is in many respects better than the sum of its parts.

M	The site visit team is very favorably impressed with the sum total of [the center's] accomplishments in all categories described above. Their use of the full spectrum of techniques to attack important questions materials has demonstrably put them in a leading position that is recognized world-wide. The published and in-progress results are impressive, and their proposal builds intelligently on the base of their accomplishments, while also describing a number of important initiatives. In sum, what we have read in the [center] proposal (and in its mail reviews) and what we have heard and seen during our site visit allow us to confidently predict that the science and technological developments coming from this group over the next few years will have a profound impact on our understanding.
N	It is the sense of the site review team that [the center] is performing fully as intended, and represents an outstanding example of what a Science and Technology Center was designed to be. [The center] is truly an interactive, collaborative and collegial group of researchers who are focused on a broad research objective, rather than a collection of independent and uncommunicative projects. We recommend that the center be renewed for the full 5 year term at an increased level of funding.

TABLE D-1 *Continued*

STC	General Comments
O	The overall impression of the review team is that this center is not only working well, but as Science and Technology Centers (STCs) were intended to work. It is clearly accomplishing more in concert than could ever have been done by individual principal investigators. We strongly recommend funding at the level requested.
P	The Center has successfully brought together a diverse group of very talented scientists to address the [center's subject area] in a broad and coherent way. The scientific projects funded by the Center are of the highest caliber and have already produced two very important results.
Q	We strongly recommend support as requested, i.e. full approval without reservations. The Center is of great value to [its scientific] community and its progress to date is excellent.
R	We strongly recommend that NSF increase funding levels in the second five-year period commensurate with expansion and enhancement of [the center's] scientific programs. The facilities of the Center are very efficiently utilized by a large number of highly interactive researchers and industrial users. We therefore conclude that NSF funds are extremely well utilized at [the center].
S	The site visitors are convinced that the core funding and stability provided by the STC have been critical to the development of a vibrant and innovative research program with impact extending far beyond the center itself. Overall, the site visit committee found this to be an outstanding STC. This center is setting an example of how we may do science as we enter the twenty first century.
T	The generation of new ideas through discussions and interactions between the multidisciplinary participants at [the center] was strongly evident to the site visit team. The extensive critical evaluation of all research projects by an interested group of peers and interacting cooperators from many different disciplines was viewed by students, postdocs and PIs alike as a strong benefit not found in University departments.
U	Support for the center has had a significant impact on both the organization of research effort and contributions made by the investigators. While some research activities were not viewed as having an impact on a level that would be expected for a center of excellence in this area, other activities clearly are at a center level and would not have been developed without the center mode of funding. The [center] has done an exemplary job in the areas of educational outreach, industrial outreach, and leveraging NSF funds both from the University, state and industrial sources. However, the ultimate success of an NSF S&T Center cannot be based on budget and outreach alone but must include the strength, innovation and leadership aspects of its research program. It is in these areas that the site visit team feels the performance of the [center] needs to be strengthened.
V	It is the opinion of the site-visit team that the STC has evolved into a unique Center for the advancement of science and technology of [its subject area] by establishing an integrated program. The STC has also integrated successfully both the educational and knowledge-transfer missions with the research programs. After consideration of all factors presented to us, the site-visit team recommends without reservation funding of this outstanding program as requested for the seventh through the eleventh years of operation.

TABLE D-1 *Continued*

STC	General Comments
W	We have no doubt that the center mode of support is justified here. The Center has concentrated its activities in areas that would be unlikely to thrive in traditional academic departments, and involves teams with members coming and going.
X	The center is clearly an excellent national resource for the fundamental study of [its subject area]. Strong and growing interdisciplinary interactions involving first-rate faculty characterize a well equipped, well planned and well managed effort with potential technological impact. The Center provides a unique environment for research and education, with major value-added to both the nature and quality of the research possible as well as to the [university's scientific specialty] community. Outreach programs to minority institutions are excellent. The only area of concern is the rather modest array of linkages to the private sector which have been forged so far. More input from and collaboration with industry would enrich the center research program and increase its impact on the private sector. The Committee recommends support as requested with the confidence that the excellent caliber of the participants, research, and management will ensure that the needed links to industry will be forged.
Y	All research projects undertaken within the Center are collaborative in nature, with at least one PI from industry and one from academia. This brings to each project investigators with different expertise, focus, and perspective. One of the criteria for Center support is that the proposed research not be suited to conventional investigator support. The thrust group structure of the Center brings together in a unifying theme excellent academic and industrial scientists to carry out timely interdisciplinary research projects in a manner that no single research group could readily achieve. The Center fully and admirably satisfies the NSF program goals for such an organization.

TABLE D-2 Comments Organized by STC Program Goals

STC	Research	Knowledge Transfer	Education and Research
A	During the first three years of the Center's development, the planning and execution of a major field experiment played a critical role in focusing research activities on important interdisciplinary problems.	The SVP recommends that [the center] strengthen the linkages between [one of] its components and other large field programs. Specifically, the SVP would like to see [the center] broaden its interaction with the [related communities] so that the future campaigns can be leveraged by [the center's] scientists.	The educational and outreach activities of [the center] take place primarily through the association with [the university]. This activity is excellent and it plays a crucial role in the education of K-12 teachers and, indirectly through the teachers, a very large number of primary and secondary school students. The SVP met with a group of about 20 postdocs, graduate students and one undergraduate. It is clear that they are all very enthusiastic about [the center]. They feel that [the center] offers them an exceptional research experience because of their close interactions with the scientists in [the center] and because of the exceptional quality and diversity of the group. In many respects the postdocs and students serve as effective bridges across the various areas of expertise in [the center].

APPENDIX D

TABLE D-2 *Continued*

STC	Research	Knowledge Transfer	Education and Research
B	Research is generally well-focused and effective in meeting center goals. The center also supports, at modest levels, high risk / high potential research. The visit team encourages the management to continue its current process of evaluation of research results to ensure that such research is directed towards center goals.	[The center] has demonstrated that it is an effective mechanism for coordinating research and promoting the exchange of ideas among productive university researchers. The program of visiting scientists provides a further link to the national and international scientific communities. The primary tie at the national level is with [a federal agency]. The primary relation to the private industrial sector is through the participation of [several] companies as investigators. Center products will impact several industries. It is desirable that the Center establish links with these industries. Progress in this direction appears imminent, because ties are being established with a number of organizations related to engineering applications which were described during the site visit.	With respect to education, a group of 45 bright and enthusiastic graduate students are now working on research under the aegis of [the center]; their contribution to the research productivity of the Center is evident. Significant efforts have been made to develop materials for public information. Due to funding shortages in the first two years, a conscious decision was made not to implement the K-12 [center specialty] program. In summary, the Site Visit Committee is satisfied by the progress [the center] has made in the broad area of outreach, and it is encouraged that grades K-12 and undergraduate education will now become a high priority program.

TABLE D-2 *Continued*

STC	Research	Knowledge Transfer	Education and Research
C	The center is very productive and the intrinsic merit of the research produced is of high quality.	The [center] has demonstrated success and continued expansion of the Industrial Outreach Program during the last year. The Industrial Affiliates Program has demonstrated a conscious effort to disseminate research results through technical meetings, annual reports, research project summaries, technology transfer days, product/process development fact sheets, collaborative research projects, and student and faculty exchanges. Participants in the IAP and visiting scientists have provided an international perspective on both the world-wide industrial needs and the current level of fundamental knowledge in the field of [the center].	While the committee is impressed with how much has been accomplished, it feels it appropriate to comment that it will be necessary throughout the term of the [center] grant that a balance be struck between education and research. We encourage the [center] faculty and staff to continuously evaluate the educational activities with the objective of maintaining congruence with the research program.

TABLE D-2 *Continued*

STC	Research	Knowledge Transfer	Education and Research
D	The research projects are of top quality, the researchers are first-rate, the education and outreach programs are extensive and impressive, and the Center is well managed. Hence, we recommend that the proposal be supported as requested. However, we believe that the impact of the Center in the next five years would be enhanced by greater integration of the research. The most visible results of the research are the standardization efforts. This work is excellent value-added from the Center. We suspect that it would not have happened without Center support; it should be continued.	[The center] has created productive informal linkages to colleagues and projects in academic institutions, government laboratories, and industry. In particular, strong connections have been built between work and the [related] industry; industrial ties have been actively sought, especially in recent years. In the academic affiliates program, the committee suggests that more structure is needed and that criteria for membership should be developed. The committee believes that, to couple [the center's] research results more effectively to industry and government applications, an exchange of personnel is needed.	The postdoc, graduate, and undergraduate education programs for women and underrepresented minority students are excellent. The outreach programs of [the center] to grades K-12 and to the community are outstanding.

TABLE D-2 *Continued*

STC	Research	Knowledge Transfer	Education and Research
E	The current report and the site visit reaffirmed that research conducted at the Center is exceptional.	Overall the team feels that to date the Center has not adequately reached out beyond itself to academic and non-academic sectors. Increasing these efforts should be a major goal in the coming years.	The Center's graduate program is very strong, with a large number of students being involved and graduated both at Masters and Ph.D. levels. Their students find ready employment in industrial positions, in established corporations and in high-tech starts-up, as well as in academia. The summer programs are innovative mechanism for reaching high school students. Although Center members are participating in a number of programs, it is still unclear what formal support they are receiving from the center and whether there are center-wide goals for these activities. It is also unclear which of these programs would not be taking place without the Center.

TABLE D-2 *Continued*

STC	Research	Knowledge Transfer	Education and Research
F	The principal researchers involved in each component field are world leaders who guide and influence the research directions of their field. Overall, there is strong evidence that the STC has added value to the research accomplishments of this group, by encouraging joint projects and by establishing an environment that has attracted new faculty members, postdoctoral fellows, students, and visitors, all of whom have contributed to the research programs. It is clear that many of the projects have been shaped or influenced by the interdisciplinary environment, and that new areas of research have opened up. The ongoing and proposed research is of excellent quality, although some areas have not been fully developed as yet. Weaknesses include the slow development of interdisciplinary research and education in [some] components and between that component and [other areas].	An institute dedicated to such an esoteric topic as [this center's] science [specialty] might not be expected to deliver practical products that are useful in the real world. The situation in the current case is quite the contrary, and in fact a multitude of interactions with industry and government are leading to saleable products and substantial information exchange. Many industrial interactions have been more along the lines of joint research.	The Center has developed two excellent interdisciplinary undergraduate programs to which undergraduate students have responded with increasing enrollment. Research topics of graduate students at the Center are all interdisciplinary in nature. Students showed enthusiasm for the cross-disciplinary opportunities that the Center provides. The Center has been energetic in its outreach programs. The program for K-12 is being actively pursued. The educational program is strong and diverse.

TABLE D-2 *Continued*

STC	Research	Knowledge Transfer	Education and Research
G	The research mission of the Center, while initially ill-defined, has in the second and third years of the Center coalesced into a major hub project. While the various component projects were not always clearly or closely interrelated, the individual aims of the majority of these projects address important problems in [the center's scientific area].	The Center has made significant efforts to make links to industry. Unfortunately efforts to date have not been very successful. The main focus of the Center does not lend itself well to interaction with industry, and thus it is difficult for the center to make linkages.	The unanimous opinion of the Site Visit Review Team was that the Center's efforts in the area of education and outreach have been outstanding. Particularly impressive is the success of the High School Teacher's Program.
H	Since the development of a [full-scale] system is extremely complex, it requires a large support group and cannot be accomplished in a short period of time. The [center] group has assembled the tools and the people necessary to examine the problem over a large number of cases and situations. This is a unique opportunity. In its 5 years, [the center] has assembled a cohesive group of scientists who have made great progress in constructing a set of powerful tools with which to attack the overall goal [of the STC]. The Center has also established an outstanding outreach program in [the center's subject area].	[The center] has already had a significant positive impact on the research community, mainly through the sharing of the [center's modeling] system. In addition, the active collaboration of [the center's] scientists with other scientists is having a mutually positive impact on the [the center's] program and other programs in collaborating universities and research institutions. In the area of transfer of knowledge and technology to non-academic sectors, [the center] is making good progress.	[The center] has done an outstanding job in their educational and outreach programs. Projects and collaborations with the Planetarium and Science Museum reach large numbers of students as well as K-12 teachers, and the general public.

TABLE D-2 *Continued*

STC	Research	Knowledge Transfer	Education and Research
I	The director and his colleagues are to be complimented on their progress in the face of the most unfortunate delay in delivery and siting of the [the center's device].	None available.	During the last year the Center has put in place an innovative outreach program. [One] is an outstanding mechanism to introduce children to science. It is suggested that thought be given to including some mechanism for evaluation of the outreach program.
J	We wish to emphasize that we agree with the previous site visit teams that [the center] has and continues to develop an excellent portfolio of programs.	The previous site visit report recommended that the leadership of [the center] promote more interdisciplinary activities and pay attention to the industrial needs of the country. The site visit team was pleased to find significant progress toward these goals. The special year on [the center's subject area] delivered what it promised. Its topic is timely and interesting. Industrial support was obtained and productive interactions with new institutions were initiated.	The [center's] education team decided to abandon the high school program in the absence of NSF support. An excellent product may disappear. Going into K-8 education may not be the most strategic use of the center's expertise.

TABLE D-2 *Continued*

STC	Research	Knowledge Transfer	Education and Research
K	This research is world class in [the center's subject area].	This STC has been proactive and creative in the development of multiple vehicles for transferring knowledge. [The center] has excelled in knowledge transfer to academia and industry. Indeed, as U.S. industry awakens to the opportunities in [the center's field], [the center] is prepared to assist them and rapidly serve their research and technology needs.	[The center] has initiated an impressive educational program that spans K-12 to post-graduate education. The K-12 outreach program is an exemplary program. More emphasis should be placed on undergraduate summer internships as a mechanism for recruiting [the center's] graduate students. More interaction should be fostered among [the center's] graduate and postdoctoral students.
L	The field of [this center] demands the interdisciplinary approach that can be obtained in an STC, and the Committee believes that this Center is now established as the strongest of all the centrally funded centers [in this specialty]. For example it has had a bigger impact than either the British or the Japanese national centers. The Center has established preeminence in several critical areas of fundamental research [in the center's specialty]. The themes and projects are well-chosen to build on collaboration between strong research programs of individual PIs, and address issues of central importance.	Actual joint work with industry still represents a small part of the total STCs effort. Industrial outreach started from a low level in 1992 and has built steadily since. We were told that only ~20% of the Center graduates during the first 5 years had gone into industry. Those students who had had industrial contacts were strongly positive and wanted more. So far there have been no or almost no industrial internships; these should be developed. In any case it is vital to provide many more opportunities for industrial interactions to the graduate students in the Center.	This Center has added significantly to existing outreach programs at [the center's institution] and has enabled this program to significantly expand its scope and impact. The particularly noteworthy activities have included a summer enrichment program for middle and high school teachers, workshops for women and minority students and an instructional van that provides important hands on laboratory experience for students at all levels of age and experience. In the center itself, some of the graduate students are particularly attracted to the interactions with the industry [in this field] and these should be further expanded. The student workshops were also universally applauded.

APPENDIX D

TABLE D-2 *Continued*

STC	Research	Knowledge Transfer	Education and Research
M	The site visit team is convinced that [the center] is without peer in the study of [its specialty]. The [center's] member institutions collectively have developed an extraordinary variety of experimental, analytical, and computational capabilities permitting research on fundamental problems in [its scientific disciplines] that would remain otherwise inaccessible. In summary, [the center] has developed a distinguished and vigorous program in [its research field] that is unmatched in scope anywhere in the world. While other groups may have comparable capabilities in specialized areas of [the research field], the site visit team concludes that [the center], due to the critical mass of scientists and facilities made possible by STC-style support, is carving a truly unique international niche into the relatively unexplored territory of [scientific] research. Their track record to date and their proposed research program hold great promise for sustained significant scholarly contributions.	From the beginning [the center] established strong linkages to other sectors. [The center's] facilities and Center-style organization allow it to provide unique services. Although [the center's] science and technology goals are more suited to providing industry with an expanded base of knowledge rather than immediate applications, [the center] has, nevertheless, developed an extensive set of links to industry.	[This] STC has been successful in its educational programs at all levels. Perhaps the most outstanding educational success [the center] has had is with their program of summer scholars. For the last two years [the center] has recruited nationally and selected a group of 6-8 undergraduates who are brought to [the university] for a period of ten weeks of intensive research project work with an assigned senior [center] mentor resulting in publication(s). This program has been notably successful in recruiting African-American students through [the center's] special connection at [a neighboring] university. The Center has initiated a significant effort in the K-12 science education area.

TABLE D-2 *Continued*

STC	Research	Knowledge Transfer	Education and Research
N	The center is formed around the efforts of a group of strong, world-recognized investigators having complementary expertise in [the center's] research [specialty]. [The center] has been particularly successful in providing a multi-disciplinary environment where students, postdocs and visitors as well as the PIs can learn each other's languages, exchange ideas and viewpoints, and develop new concepts. The research is first rate and highly productive, and represents cutting edge approaches to high risk, high payoff objectives. We were all impressed with the strong research interactions between the participants over a range of disciplines.	[The center] has obviously taken the recommendations of the previous site review committee to heart, in that they have made a very conscious and conscientious effort to expand and improve their knowledge transfer with the research sector. [The center] has developed a large number of linkages to other universities in the US, primarily through collaborative research efforts between individual faculty members of those universities and members (or facilities) of [the center]. [The center] has been especially active in forging links with some of the strongest research groups in [its specialty] in other countries. [The center] also has a number of significant government laboratory interactions. There are numerous linkages to industry.	The site review team was extremely impressed with this aspect of [this center]. The commitment of the entire [center] team to this area is obvious and touching. Several of the [center] programs are primarily directed at improving the science and technology education of students in grades K-12 in the [local] area. The researchers and the students presented strong arguments that this program has been successful. The review was impressed by the diverse cross section of students who were represented and by the anecdotal stories of the successes of the students who have participated in the program thus far.
O	The research participants in the [center's] project are outstanding and clearly capable of achieving the goals set forth in the proposal.	The Center has made a consistent effort to build strong ties to the local community and the technical community at large.	[The center] has created a model Education and Outreach program at the K-12 level. At the university level, significant numbers of undergraduates and graduate students have been able to participate directly in the [center's] project.

APPENDIX D

TABLE D-2 *Continued*

STC	Research	Knowledge Transfer	Education and Research
P	The visiting committee unanimously agreed that the scientific efforts of the Center are of the very highest quality. These projects provide strong examples of how the Center enhances the scientific impact of NSF funding by creating a whole that is greater than the sum of its parts.	The unique requirements of [this center's specialty] usually translate into a market that is too small to be of much commercial interest; thus, the [center] might be excused if technology transfer, in the sense of the movement of patented or potentially patentable devices or technology from their university birthplace to some industry where they could be commercially exploited, occupies a minor position in its list of accomplishments. The Center has had some modest successes in this area and has initiated programs that promise more.	The [center] should be commended for its attempts to serve as an educational resource in both the scientific and lay communities. Through [one] program, the [center] has sought to improve the working lives of its members. And through its varied educational "outreach" programs, the [center] has shared its human capital with amateur [scientists], local schoolchildren, and other groups who would not ordinarily have access to cutting-edge research. The committee was tremendously impressed with the involvement and commitment of several people in the wide array of [center]-related educational programs, including local speaking engagements, mentoring, internships, and laboratory apprenticeships.
Q	The strengths of the Center include: – strong cross-disciplinary research projects – the development of new technology that is driven by the scientific programs – high quality faculty with funded research programs and bright graduate students.	The STC has followed its declared strategy and set an enviable example in successful academic entrepreneurship for each of its programs, by developing an array of ties with institutions in each of the areas mentioned above. [An industry laboratory] and the STC have initiated negotiations to construct a [device].	It is clear that the educational opportunities associated with the Center are unique and are having a significant impact not only on student recruiting but also on the development of these students into well trained and enthusiastic scientists. Several outreach programs have also been effectively implemented.

TABLE D-2 *Continued*

STC	Research	Knowledge Transfer	Education and Research
R	The [Center] has established itself as a global leader in the area of [the center's specialty]. The Center has achieved substantial visibility in [its specialty]. The technological developments pioneered within the Center have been enormously influential throughout the community.	The Center has developed numerous linkages to outside sectors. Particularly impressive has been the interaction with industrial concerns. The Center has been involved with several companies in [its state], including spin-offs of the Center in the [local] area, as well as some other U.S. companies outside of [the state] and foreign companies.	The review panel found that the efforts of the [center] in all aspects of education were excellent. Support of graduate education by the [center] has recently been augmented by the additional award of [several] NSF Graduate Traineeships. The K-12 outreach and science education programs were thought to be well-targeted and much needed.
S	The projects are at different stages of maturity, as expected for a program entering its seventh year. The outstanding track record of STC in its mature projects, as well as in projects from earlier years, provides confidence that good judgment will be used in guiding the newer ones forward. The STC is a definite leader in [its specialty] field and, as a total program, has very few, if any, peers. While the research and development work of this STC is in some cases not as unique or advanced compared to the state of the art as it was in the Center's early years, this is in large part attributable to the success of this Center in distributing its ideas and technologies.	The STC has maintained an excellent record of collaborative interactions with industry. These involve not just the PI, but many of the senior people. The companies involved range from very small [firms] to multinationals, and span a considerable geographic range.	The interaction of the STC with the surrounding community goes beyond the missions of the individual scientists doing basic research and development. The numerous educational programs extend into virtually every part of society and place special emphasis on school levels K-12. It was very clear that the members of the department view the evolving outreach program with enthusiasm.

TABLE D-2 *Continued*

STC	Research	Knowledge Transfer	Education and Research
T	The site visit team know of no other group worldwide pursuing this course in such depth as [this center]. Although [the center] has been in operation for a very short time, significant accomplishments have been made in each of [its several] program thrust areas. There are excellent prospects that work in progress will result in significant advances in the near future and that these will result in seminal publications in prestigious scientific journals. Current research on [activities in a large number of subspecialties] may spread the effort of the team too widely. The emphasis on multi-disciplinary, long-term goals in a Center mode has led to a relatively small number of publications in peer-reviewed publications.	[The center] has developed an innovative mechanism for cooperation and knowledge transfer with industrial corporations. Companies that have a substantial research capability in [the center's specialty] can become associates of [the center]. The value of [the center] as a potential resource has been recognized by companies [working in this center's specialty field].	In summary on the basis of various indicators available to assess quality and effectiveness of similar programs, the site visit team concluded that the outreach and education program had exceeded expectations and has a high potential for continued major accomplishments. The outreach program to train inexperienced high school students in laboratories may be too time-demanding on researchers for the return to the program.

TABLE D-2 *Continued*

STC	Research	Knowledge Transfer	Education and Research
U	In research, i.e., the development of new principles, new knowledge generation, rigorous theory testing, etc., the record of the [Center] during the last five years is mixed. The quality of individual research projects is uneven and this is true within and among all of the research thrust groups. While [the center's] activities have contributed to the development of knowledge and techniques for studying [this center's specialty], they have not yet resulted in facilities and/or resources that would be sought out by the national or international research community.	The quality and quantity of knowledge and technology transfer from the [center] to non-academic sectors in the area of [the center's specialty] is very promising. The relationships which have developed with the state and some industries, though in their early stages, appear to have excellent potential. The hurdles already crossed and the field trials to be done are truly impressive. The existing linkages of the [center], which fall into several categories, are generally excellent.	One of the brightest aspects of the [center] is its educational program. The K-12 program involves local, regional, and national efforts. These efforts have had a very positive impact on the scientific literacy of students [in the center's specialty]. It is apparent that the [center] has influenced the curricula for K-12 through training programs for secondary school teachers and students. The undergraduate educational program has developed strong linkages with the faculty and students from [several] campuses of [a university with primarily minority students]. The graduate education program is of high quality, broad in perspective, encourages multi-disciplinary interaction and appears to be well integrated with the Ph.D. programs of participating departments.

APPENDIX D

TABLE D-2 *Continued*

STC	Research	Knowledge Transfer	Education and Research
V	[One specialty] is successful in bringing new thoughts towards achieving the objectives for the STC. The exceptional number of publications and presentations is a clear manifestation of this success. The principle investigators are presenting their work on the correct venues throughout the world, and they thoughtfully have chosen the proper journals to publish their work. The quality of the individual research plans and the uniqueness of this STC's research contributions are well recognized. It is the opinion of the site-visit team that the STC has created a unique research theme and environment within communities. The uniqueness arises from the successful joining of [scientists] and engineers in [conducting experiments in the center's specialty]. The impact of research contributions in the areas of [the center's specialty] has been very strong, with the principal investigators establishing both an outstanding record of scholarly work and a high degree of credibility in the industrial world. The overall quality of publications is very good, with a significant number of publications being outstanding contributions to the literature.	The STC has done an exemplary job in the area of knowledge transfer. In the course of its research is has interacted with [many] industrial organizations in addition to several federal laboratories, research institutes, and universities. This intensity of interaction is well in excess of what one might expect from a university research gram; however, it is well in tune with the defined mission of a STC. This degree of industrial interaction may provoke concern about damage to the main mission of the university, namely, the production of well educated graduates. However, this STC appears to have reached a good balance between education and the potential constraints imposed by more targeted research. In fact, the quality of the graduates has been enhanced from an industrial point of view. Discussions with a representative group of external advisory board members indicate that the technology transfer to industry is highly regarded and that programs reaching this stage of completion have been relevant to the needs of the industrial partners.	Graduate education stresses genuine interdisciplinary exchange, a point well documented by the graduate students currently in the program. They know of each other's work; they learn from each other; they feel privileged to be part of the program. Communication skills are stressed throughout their career, more so than most institutions. The site-visit team believes the STC is preparing students well for their future.

TABLE D-2 *Continued*

STC	Research	Knowledge Transfer	Education and Research
W	The utility of these tools for research is attested to by a substantial number of books and publications in leading [scientific] journals [in the center's specialty] based on work stimulated by these tools and, more generally, work done at the Center. Nevertheless, viewed merely as an investment, the Center has not yet paid off as well as we might have liked. By the time of the next visit the Center should be able to distinguish between results achieved mainly at the Center and results obtained outside of the Center proper but using tools developed at the Center.	This should hardly need saying, but let us also note that [Center activities] tend to be too specialized for commercial sales. Indeed, any attempt at commercialization, even though it would bring in small revenue, would hinder the main mission of getting the tools widely used in the [its scientific] community.	As has been reflected consistently in previous site visit reports, the educational programs, particularly on the K-12 levels, show early signs of being highly effective. In the university context, the Center's vision for educational outreach and the incorporation of [its scientific specialty] is tremendously exciting and highly ambitious.
X	The quality of the individual Center research participants is excellent. The productivity of the participants as gauged by publications in high-quality journals is excellent. Many of these evidence strong and growing multi-investigators, multi-disciplinary, and multi-institutional interactions—a crucial aspect of STC research. Despite the high scientific merit of the research accomplished to date, the influence of the private sector on this effort has only been modest, and it is fair to say that the results to date are unlikely to have any major impact [related to the center].	Several linkages and collaborations with external laboratories have been established, and others will be in place in the near future. In general, closer ties and more collaborations with industry are needed. Many aspects of the STC could benefit from improved "industrial outreach." We applaud the Center for what they have accomplished in this arena so far, but feel that additional effort should be expended along these lines.	The Center has met its primary goals with regard to graduate education. After an initial slow start, the Center has made impressive progress in putting in place effective and sustainable programs to reach outside the local university community. The Center has done a particularly impressive job of establishing summer programs for middle and high school students. The Center is making satisfactory progress toward establishing effective and sustainable programs to reach outside the university community and to encourage underrepresented minorities to enter science and engineering fields.

TABLE D-2 *Continued*

STC	Research	Knowledge Transfer	Education and Research
Y	The Site Visit Committee was uniformly impressed with the consistently high quality of the research projects that form the main focus of the [Center]. It is clear that the STC has already matured into a world-class force in this area, with an array of distinct and interrelated projects presently ongoing that collectively form a basis for future advances in the [center's specialty]. Both the quantity and quality of the research accomplishments throughout the Center are extremely good to truly outstanding. Some of the ongoing projects have counterparts in other institutions. However, the uniqueness of the research contributions of the Center derives from the consistent synergy demonstrated between the academic and industrial collaborators at the individual project, thrust group, and overall Center levels.	The [center] is a unique enterprise because of the shared scientific interest of [several] partners and the special opportunities for interaction and sharing of resources made possible by their proximity. The establishment and nurturing of [the center] has been an earnest and dedicated effort on the part of the [center's partners].	A number of Center educational accomplishments are noteworthy; the contributions of the industrial PIs as teachers and mentors for graduate students and postdocs; the enriching role of visiting scientists in the Center; effective programs for high school and college teachers; and the first-class annual symposia sponsored by the Center. The philosophy of the education and outreach programs—to use basic research as a teaching environment—serves everyone well.